Structures and Morphisms

Structures and Morphisms

Published by Amarilli Books

Copyright © 2017, Brian Capleton PhD

briancapleton.com

First Edition 2017

ISBN 978-0-9935372-4-0

A CIP catalogue record for this book is
available from the British Library.

Cover illustration: 'Parklands', acrylics on
canvas, by Brian Capleton

Mathematically, in the right circumstances, anything can undergo a morphism into anything else.

Contents

Preface

Mathematics

The most interesting thing about applied mathematical functions does not lay in how they can describe natural phenomena, but when they cannot. Singularities in functions are points where they fail to describe the phenomena of nature, with some arguable exceptions such as the phenomenon of the black hole, and the modern cosmological view of the origin of the universe. But in these places is also the most interesting thing about natural phenomena itself, as understood through the human brain and the intelligence arising there.

Mathematics exists in our brain, in our mind, as much as it exists in nature, through which our brain, our mind, arises. Modern science can see that all of nature is mathematics in action. There is mathematics in life forms, there is mathematics in art. It is in the sciagraphy, the perspective, the illusion, and in the very beauty in all the arts. It is in the form and substance of music, and even in the manifestation of genius in art and science.

There is mathematics in nature and in the human brain that nature has produced. The appearance, the experience of the music of colour and form, and meaning in all the arts, is all created through mathematical form in action. All life on Earth, and all our conditional human experience of being, is the manifestation of this mostly hidden mathematics.

There is mathematical form in the complexity of the greater network of all life forms on Earth. You can't really understand the relations between forests, or between forests and deforestation, just by understanding the internal chemistry of leaves. Those relations grow out of the internal chemistry of leaves, but not *only* out of that. They grow out of the network of everything, everything that we need to come to understand the nuts and bolts of, like the internal chemistry of leaves.

Before we can put that knowledge to its fully realised use, with maximum benevolence, we have to comprehend the nature of these other relations. We do need to understand the internal chemistry of leaves, and the squares on the hypotenuses, but unless we see the bigger picture, we don't really understand our situation. For that understanding to happen we need to see the secret behind the appearance of mathematics itself. The secret behind the manifestation of data.

Abstract mathematical form is not an ultimate transcendental reality behind the scenes of our existence. It is part of our existence. It is a bridge between our understanding of our own mind in science, which we must rise to, and our understanding of our existence in terms of the mechanics of nature, which gives birth to our own mind. Existence exists because it is a possibility. What we are, as human beings, is the exploration of this existence - the exploration of the possibility of our experience of being.

Today we know in physics that uncertainty is fundamental to nature, and that concreteness is ultimately an illusion. Traditional mathematicians loved the idea of the concrete and certain. They loved the idea of absolute truth in mathematical structure. The satisfaction that the square on the hypotenuse always equals the sum of the squares on the other two sides. Except that it doesn't, necessarily, depending, for example, on the surface.

Logical positivism came to an end, just as absolute faith in an absolute set theory, or in any system of understanding of mathematical structures, had the carpet pulled from under the feet of the believers by those who saw through it, like Russell and Gödel. There is no absolute truth in mathematical structure.

Today, the faith in absolute, objective concretes and proofs, as the domain of the deepest truths, is only something that can still be clung to, by those who see with blinkered vision, and aren't really paying attention to the 21st century world as a whole.

Artists generally "speak a different language" to mathematicians, but both are using the same generic means of speaking - the human brain. If we knew the human brain better, if we knew the human mind better, we might not see such a distinction, or see it in the same way.

The traditional *modus operandi* of many mathematicians has been faith in concrete reality and truth outside the self that seeks it. The *modus operandi* of the greatest artists has always been the communication of aspects of a truth that is only discoverable by knowing the mind and self of the one who is looking, hearing, speaking, or creating in any way.

When we truly realise that it is the brain that manifests both the artists and the mathematicians, and that the mind is a creation of nature through a principle of the brain not yet grasped by science, then we might begin to see that art is the manifestation of an as yet undiscovered mathematics, and that there is a mathematics that is an art.

Structures and sets

No matter what you believe about the world, whether you believe it is under the dominion of politics, love, God, or human nature, facts are facts. And it is in the grasping of facts, beyond opinion, beyond personal belief, that the potential and ability of science to manifest practical compassion for human beings, lays. The facts about nature are the facts about nature, beyond anyone's personal beliefs, and it is the business of science to discover those facts. However, we are doing this without knowing very deeply, what a fact really is, and where it comes from.

Our ideas of *cause and effect*, as where things come from, is usually rooted in our idea of time. Few come to see that time itself, the apparent flow of cause and effect as time, has a cause, beyond time. For the first time in science there now exists a concept for this, dressed in such ideas as quantum correlations, and mathematical quantum structures. But all this still rests on our idea of mathematical facts.

Our grasp of mathematical facts is something that is happening in our intelligence. It is not such a large step to see that the cause of this thing we perceive and see in the phenomena of our world, that we call mathematical fact, lies in the nature of our own intelligence, the intelligence arising in nature, that we are being. And it is only another small step to see that our mathematical understanding of nature is the beginning of the scientific understanding of the nature of our own intelligence.

Sets as we conceive and understand them are not a mathematical fact about anything separate from our own mind. Rather, they are an interpretation of mathematical facts, made in the state of our own intelligence. The concept of a set is a concept. That means, it is something happening in our mind, and in the functioning of the neural networks in the human brain. These networks are structures, and their functioning is according to mathematical structures, mathematical facts that are a part of our world, and our brain functioning. The idea of a set is part of those structures.

So to get to the root of it all, to get to the root of our own intelligence, to get to the root of the intelligence we are being, we will have to go beyond the idea of a set, as the basis of mathematics, into the understanding of these structures. And that cannot be easy. It cannot be easy because the idea of a set as something outside our own mind, no matter how objective or object-oriented we are being, rests on a certain ignorance of who we ourselves are.

About *Structures and Morphisms*

Mathematics is an object-oriented science and philosophy. And so it is that infinity in mathematics is an object. It is an object that is distinct from number-objects. Infinity appears in mathematics either in the form of a process, and/or in the form of the escape of an object from identification as a number-object. In both cases there is an "escape to infinity", a term that will be familiar to those who work with certain fractal mathematical forms.

Where there is a mathematical function, or a morphism of any kind, there is a domain D and codomain C. In structure theory, as in many other mathematical approaches, we would say $D \xrightarrow{f} C$. In structure theory f can be a *structural* morphism, which is ultimately a change or relation between structures that have been created from a *bounded infinity*.

The theory in *Structures and Morphisms* arises from the analysis of how relations between distinct but identical objects arise. Apparently distinct and non-identical objects can then arise as structures out of the former. The role of infinity as an object in mathematics then turns out to be central to these structures.

Distinct but identical objects appear in equations as variables and constants, and in physics as fundamental particles. There are circumstances in the understanding of any structure of relations between objects, in which the distinction between identical objects disappears. One of the first places we see this happening is in the infinite singularity, the object of infinity. There, we can see that the dualities of infinity that appear in mathematical structures are artefacts of structure interpretation.

In *Structures and Morphisms* infinity is discovered not as an anomaly or inconvenience in the nature of ordinary functions, but as a very root cause behind functions themselves. Through the object of *bounded infinity* even unbounded infinity can be handled through the objects of the *i-ratio* and *i-product*. Even finite numbers can be viewed as finite transformations of *i-numbers* and their ratios.

Structure theory

The whole of this volume could be considered as a kind of tutorial in the basic ideas of structure theory. Nevertheless, the following condensed description superficially covers the essence of the theory and may make the rest of the contents easier to follow from the beginning, which otherwise may seem a little abstract to begin with. The ideas laid out in the main volume provide some material for further exploration of the relation between mathematical form, nature, and our own intelligence, which one day, perhaps sooner then we think, as our knowledge of the human brain grows, will have to take us beyond our current naive realism.

Structure theory deals with bounded infinity. Perhaps the simplest way to envisage a bounded infinity is as the Euclidean points of the circumference of a circle. When the nature of bounded infinity is properly understood it can be seen that the concept of unbounded infinity can be handled in terms of bounded infinity.

Structure theory treats the native form of numbers in nature as numbers to the infinite number base. In the infinite number base there is no need for numbers constructed in columns, or arrays (which are constructs of convenience made in our intelligence), as there is already an infinite theoretical supply of unique number names, symbols, or simply unique number-objects. Conversely, in the unary number base there is only one number object, that has to be aligned in an infinite array or "columns", by instantiating the same number object over and over.

Thus, in the unary number base, unique numbers are constructed as arrays made from multiple or infinite instantiations of identical objects. The structure of numbers greater than unity is fixed by the structure of the array. Conversely, in the infinite number base, there is an infinite set of *unique* number-objects, and numbers greater than unity are the structures of relations in the infinite set of these unique objects.

These two schemes correlate respectively, to the relation between what structure theory terms *structure space*, and the ordinary concept of a number space such as $\mathbb{R}$, $\mathbb{C}$, or $\mathbb{N}$. An infinite Hamiltonian cycle through the objects of a structure space, or a space of unary numbers, constitutes a bounded infinity. However, the infinity of objects in this structure space also have a full network of relations that is identical to that of an infinite number-base, if its objects were to become unique.

In this way, number-space can map onto structure space, in the way that numbers to the infinite number-base can map onto numbers in the unary number-base. But conversely, structure space can map onto infinite natural number-space (to the infinite number-base) in infinitely many different ways.

When the nature of the relation between the continuum and discrete numbers is taken into account, the mappings between structure-space and number-space (either to a finite number-base or to the infinite number-base) can be accounted for in terms of bounded infinity considered as a continuum, and all structures can be regarded as structural functions of the bounded infinity.

When we view the nature of the world or any system, including mathematical structures, in terms of structures of relations between non-identical elements of any kind, then these structures are a "mixture" of "finite portions" of these two kinds of spaces. Systems, or networks of relations between objects, can consist of many different kinds of objects, but in structure theory any object can be considered a structure, and any structure, as a network of relations, can be considered an object.

Beyond the theory of everything

One view sometimes found in the field of modern science is that we are looking for an ultimate, definitive understanding of the nature of the universe, and of our own ultimate origins - who we are, and where we come from - in terms of objects and functional relations between them. This is sometimes referred to as the search for a "theory of everything".

No such understand can exist, because all such understanding of objects and their relations amounts itself to dynamic structures of relations in the neural networks of the human brain. Only in the understanding of how and why the objective phenomenon of the brain amounts to the experience of mind and being, can any true understanding begin. In fact, the very idea of a proper understanding of the nature of our world, and of who we are and where we come from, in ignorance of that, is just naive realism. Structure theory is not only about pure mathematics or the philosophy of mathematics, it is mathematics in the context of post naive realism.

Because structure theory itself deals with *objects* and their relations, and is object-oriented, it might at first sight seem to be itself an expression of naive realism. In fact, it is the opposite. It deals with objects and structures of relation between objects, because such structures are the basis of material cause and effect, and of our scientific understanding, and as such, are indispensable. The ideas in this volume are no more than a pointer towards a post naive mathematics, as an embracing context of all object-oriented mathematics, and what it says about the nature of the phenomena of our world.

Objects and Structures - Beginnings

Two objects are either distinct, or behind the appearance of their duality, they are one and the same thing. If they are one and the same thing, then the duality is an illusion or an artefact of our description of the object.

If they are identical, we may informally say they are the "same object", but this is merely a colloquialism. Truly identical but distinct objects are always separate *instances* of one object.

We can define an *object* as any named thing under study. We define a *structure* as a network of different objects and all the relations between them. We can also call any structure an object, and consider any object as a structure consisting of one or more objects, as necessary.

In structure theory objects can be *declared*, or *specified*, without necessarily knowing or specifying any defining structure or further properties. If we say there is an object **O** , when we have *instantiated* the object, but have not specified anything else about it. We have also instantiated the written symbol, standing for the object, which is usually distinct from the object it stands for. However, we can still refer to this symbol in discussion as "being" the object it stands for, if this is appropriately possible, as long as our meaning is understood.

We have already introduced a *structure* of the *relation* between the symbol and the object it stands for. Furthermore, this is not the whole extent of the structure. The structure extends to include our own mind and intelligence, in which we have made these instantiations through thoughts. And there, the structure extends to include the neural activity of the brain, through which this happens.

Structures are defined by *specifying* relations between objects, as *intervals*, even if the nature of those relations is unknown. Intervals can be depicted by double-headed arrows ↔ between object symbols, a wider variety of which will be discussed shortly.

A *full structure* has the form of a fully connected graph in which the objects are the nodes, and the relations between the objects are the graph edges, which are the intervals. Any multiple edges are considered already embodied in each interval, according to specification, and loops are considered as part of the nodal object, and ignored, unless otherwise discussed.

We may often refer to the full structure in a given context simply as *the structure*. Any *part* of the full structure can also be referred to as a *structure*, but formally it is a *partial structure*. It is a *partial representation* of a larger structure, showing only some of the intervals. Much discussion can take place around *partial structures*, ignoring many intervals. This is possible according to the context of the discussion. So when it is necessary to refer to the whole graph, then if there is any doubt, the term *full structure* is used.

Intervals are objects or structures whose meaning in a structure is twofold as (a) a relation between objects or structures that they connect, and (b) the designation of a "difference" between objects that they connect, that makes them distinguishable.

Structural "difference" is what makes objects (and structures) *distinct* and specifies that they are *not one and the same object* (structure). We use the double-headed arrows $\leftrightarrow$ to signify that distinction, e.g. $\mathbf{A} \leftrightarrow \mathbf{B}$. If two objects are distinct, then there is by definition a *relation* between them, even if they are identical, and hence the interval specifies the existence of both the structural "difference" *and* the relation.

Two identical objects can only be distinct, rather than one and the same object, if there is a third object to which each is related differently in some way. If they are distinct but *identical objects*, then the interval between them contains the fact that each has a different relation to a third object or structure, which may be a *space* or another object. So the interval in the first instance represents this third object (which of course may be itself a structure of many objects):

$$\mathbf{A} \leftrightarrow \mathbf{A} \equiv \mathbf{A} \leftrightarrow \mathbf{S} \leftrightarrow \mathbf{A} \quad (\mathbf{S} \text{ is a space})$$

or

$$\mathbf{A} \leftrightarrow \mathbf{A} \equiv \mathbf{A} \leftrightarrow \mathbf{O} \leftrightarrow \mathbf{A} \quad (\mathbf{O} \text{ is an object})$$

A space as in the first example, is itself an object. It is an object with a specified name - *space*. This may carry with it additional implied specifications, according to context.

So the arrow interval $\leftrightarrow$ between two objects represents another object or more structure, and is "shorthand" for the way this additional structure "connects" the two objects with a *relation*. In structure specifications whether or not an object (structure) is regarded as a *space*, is a question of the context and *meaning* of the structure. To talk about two particles in space, for example, is to instantiate three objects, two particle objects, and one space object.

To reiterate, intervals are themselves objects, so any structure $\mathbf{A} \leftrightarrow \mathbf{B}$ may be expanded to the form $\mathbf{A} \leftrightarrow \mathbf{C} \leftrightarrow \mathbf{B}$, whereupon the new intervals can also be treated in the same way. This is a process that can be carried out infinitely.

If we did this with the example of the two particles in space, we would see a structure $\mathbf{A} \leftrightarrow \alpha \leftrightarrow \mathbf{C} \leftrightarrow \beta \leftrightarrow \mathbf{B}$, where $\mathbf{C}$ is the space, and α and β are relations of the particles to that space, for example, position in that space. Carrying out the procedure again we would see

$$\mathbf{A} \leftrightarrow R_1 \leftrightarrow \alpha \leftrightarrow R_2 \leftrightarrow \mathbf{C} \leftrightarrow R_3 \leftrightarrow \beta \leftrightarrow R_4 \leftrightarrow \mathbf{B}$$

where for example, if the α and β are positions in the space, then the Rs might be the relation of some property of the particle to its position in the space. Or they might be any other concept or part of the structure of our understanding of what we are talking about when we speak of "particle" and "position" and "space", and so forth, and the relations between them.

The structure shown is a partial structure of relations between the objects. In the full structure there will be many more intervals, waiting to be replaced with more objects and structure, that we haven't mentioned here. Many of these are just taken for granted when we use words like "particle" and "space", following from our education on what these things mean, and are contained in the psychological apperception mass for the concept.

Ultimately, all concepts and thoughts that we use to understand the objective structure or system in question, are part of a network of meaning that is itself a structure. A scientific paradigm, for example, is such a structure. And in terms of neuroscience, the whole network of meaning is in the structure of the functioning of the brain.

More examples - structure space and infinity

We can think of the relationship between objects and intervals in the following way. In structures, a *relation* (interval) between two objects is something that is often already *expressed* in terms of the two objects, but still stands to be "explained" in terms of further structure. This is analogous to the concept in physics, of action at a distance through an intervening field. We can express the gravity force, for example, in terms of two masses at a distance, but there is a field that intervenes.

The numbers 3 and 4, in a pure mathematical example, differ by the unit of the two numbers, but there are other aspects of their relation, some of which imply a continuum structure between the two numbers. This arithmetical difference is true of any two adjacent integers. However, each pair of adjacent integers also has a unique relation of *ratio*.

So although the conventional arithmetical "interval" as it is called, between each pair of adjacent natural numbers is the same, the structural *interval* (arrow) between them, one aspect of which is the arithmetical ratio, is unique between each pair. This is a specific simple example of how two distinct non-identical objects differ, *and* have various relations between them, which structure theory expresses through the interval arrow.

In the structure $1 \leftrightarrow 2 \leftrightarrow 3$ the intervals are unique. In the structure $2 \leftrightarrow 2 \leftrightarrow 2$ the intervals are identical. The number objects here are identical, too. In any set of identical objects *without further specification*, the intervals are identical. Because the intervals themselves are objects (structures), we can expand this to $2 \leftrightarrow \alpha \leftrightarrow 2 \leftrightarrow \alpha \leftrightarrow 2$. And because we can do this infinitely, the intervals in $2 \leftrightarrow 2 \leftrightarrow 2$ can be regarded as infinite fractal *spaces* composed of *interval-objects* that are identical.

So what distinguishes the number-objects here, making them distinct, is this infinite fractal *structure space*. There can also be another structure, $2 \leftrightarrow 2 \leftrightarrow 2$, that is distinct from the first one. If we write the first structure as $(2,2,2)$, and the second structure as $(2,2,2)$, then without further specification, we don't know whether these two structures are distinct or not. If however, for example, we wrote

$$D_1 \leftrightarrow (2,2,2) \leftrightarrow (2,2,2) \leftrightarrow D_2$$

and specified that D_1 and D_2 are two distinct methods of obtaining the data contained in the two objects in the parentheses, then we have specified that those two objects are distinct. They are two distinct sets of data obtained in different ways. We could alternatively just write, for example, $(2,2,2) \xleftrightarrow{D} (2,2,2)$.

Any interval like this, between two identical objects, stands for further structure, in the context of which the two objects are distinct. This further structure, in structure theory, is a *space*, although here it is not *structure space*, because it has been specified. If there is no such structure specified, then the interval between two identical objects is *structure space*, essentially, still waiting for further specification.

Mathematically, in isolation, these two objects are not distinct at all. They are both the same set of numbers. They are two *instantiations* of one and the same object. In the structure of their instantiation here, they are distinct. But they are also one and the same object that has been instantiated twice.

Out of the context of further structure, every number is a unique object. Wherever it appears in a structure, it is one and the same number, and yet in the different instances of it in the structures, it is not the same, but distinct.

In precisely the same way, the object ∞, or infinity, in mathematics, or if you subscribe to set theory, a given cardinality of infinity, $\infty_\aleph$, is one and the same object that appears in distinct instances, in distinct structures.

The word *number* usually tacitly refers to a finite number. The term *transfinite number* may incorporate the word *number*, but a transfinite number is a distinct object from any *number* in the usual sense of the word. "The set of all finite numbers" is itself an object. It is an object that is commonly mathematically regarded as "an infinity". Let's call it ∞_F for brevity. If we instantiate a structure **S**, as $\mathbf{S} \equiv \infty_F \xleftrightarrow{\alpha} \infty_F$, then we also have by definition that the structural interval α is the set of all relations $n \leftrightarrow n$ where n is any number in ∞_F. In other words, α is none other than the *structure* of the set ∞_F of all finite numbers, and **S** is that structure's relation to itself.

In the expanded form $\infty_F \leftrightarrow \alpha \leftrightarrow \infty_F$, the two intervals are the same. Expanding again as

$$\infty_F \leftrightarrow \beta \leftrightarrow \alpha \leftrightarrow \beta \leftrightarrow \infty_F,$$

then β is the relation of ∞_F to its own structure. The relation of an object to its own structure is one of identity. In other words, the structure of ∞_F is such that ∞_F can be instantiated twice, or indeed multiple times, and all instantiations will refer to one and the same object.

The structure $n \leftrightarrow n$ can similarly be expanded to $n \leftrightarrow \gamma \leftrightarrow n$ but here, γ cannot be seen to be any particular set, or to be the structure of n, unless we regard n as its arithmetical continuum interval from zero, in $\mathbb{R}$ or $\mathbb{C}$. Then γ is the set of all relations $m \leftrightarrow m$ where m is any number in the arithmetical interval continuum $0 \rightarrow n$. The interval $0 \rightarrow n$ is also an infinity. So we also have that for any number n when considered as an arithmetical continuum interval from zero, in $\mathbb{R}$ or $\mathbb{C}$, all instantiations of the same number will refer to one and the same object.

We thus see that structurally, any number in $\mathbb{R}$ or $\mathbb{C}$ is a single object that can instantiated multiple times in mathematical structures. These mathematical structures are not the structures of symbols through which we express mathematical structures, such as equation or written functions. There is also a structural relation between a written mathematical structure W and the structure **S** that it refers to. An algebraic structure W is a structure composed of symbols, that symbolises a structure **S**, that in turn is composed objects that are number-spaces related by structural intervals.

We can see a very simple example of multiple instantiation of a single object, in the expression $2x + 2y$, where the structural reason we can equate this to $2(x + y)$, where 2 is one object and $(x + y)$ is another, is that the two instances of the number 2 can be "collapsed" here into one instance. We can do the same thing with x and y if it turns out that they are equal, and therefore two instances of the same number.

Ordinary written mathematical functions can therefore be viewed as structures in which the algebraic objects are nodes in a structure with relations or edges between them. However, this is not necessarily the same structure as the structure of number-spaces to which it refers. The former is literally a translation of the latter, in the "language" of the mathematical mode of expression that we have developed, which itself consists of structures of objects that can be written as symbols (including number symbols).

These structures, in this "language", are object-oriented. But they do not contain the objects to which they refer. The rules for the manipulation of these structures, that we learn as mathematics, requires our own intelligence, and they become, in use, an expression of our own thought as structured and disciplined by this learning. So there is a relation between these structures and our own intelligence, and even this we can write as the structure $W \leftrightarrow \mathbf{S}_q$, where $\mathbf{S}_q$ is the structure of our intelligence.

There is also a relation between the structure $W \leftrightarrow \mathbf{S}_q$ and the phenomena of nature, because the phenomena of nature is "governed" by mathematical structures. Even in the simple expression $5 = 3 + 2$ is the fact that 3 material objects and 2 material objects are together, 5 material objects. But the numbers, and the understanding of this in terms of numbers, is part of our intelligence.

It is now well known that many non-human animals can recognise specific quantities. The use of counting using a number-base system, and the concept of numbers, however, is a feature of human intelligence, and even if it were found in another species, it would still be a feature of intelligence. It is a feature of *brain functioning*.

Specific quantities are also "recognised", as it were, in systems in nature, governed by mathematical structures, that do not constitute a brain organ. Chemical reactions, for example, are based on valency which is based on the relations of the *numbers* of electrons in the outer shells of the atoms. Our mathematical intelligence is an expression of nature and nature's laws. It does not follow from this, however, that in using this intelligence in the object-oriented way, that we are exploring structure, or objects, that are other than the intelligence we are being. For this reason, a little later, we will be looking again at the role of $\mathbf{S}_q$ in structures.

Multiple meanings of interval arrows

So two objects or structures **A** and **B**, and the relation between them, can be represented in its simplest form as $\mathbf{A} \leftrightarrow \mathbf{B}$ (where the double ended arrow $\leftrightarrow$ is again the interval). A single arrow may stand for one or many relations between objects or structures. If we say $\mathbf{A} \leftrightarrow \mathbf{B}$ there may be many intervals from objects in **A** to objects in **B**.

Dualities or multiplicities of identical objects are distinct, different *instantiations* of the same object. In the structure $\mathbf{A} \leftrightarrow \mathbf{A}$, the two objects are separate *instantiations* of the object **A**, each instantiated object being identical to the other.

An interval is a relation between adjacent objects in the structure graph, that as such may embody as many different relations between the adjacent objects, as we can specify or find. Interval arrows may be augmented as, for example, $\xleftrightarrow{\alpha}$ or $\xleftrightarrow[\beta]{\alpha}$, as necessary, in order to convey more specification. In the event that two written object symbols actually refer to one and the same instantiation, then the arrow is written $\xleftrightarrow{\leftrightarrow}$, or $\nleftrightarrow$, or the $\equiv$ sign is used.

Redundancy

As we have said, a structure of $\mathbf{A} \leftrightarrow \mathbf{B}$ may be expanded to the form $\mathbf{A} \leftrightarrow \mathbf{C} \leftrightarrow \mathbf{B}$, but this is not necessarily always fruitful or even meaningful. The expanded form now has 3 intervals in the full structure, but we might regard the third, $\mathbf{A} \leftrightarrow \mathbf{B}$, as now redundant. In many cases intervals can be treated as redundant, for the purposes of the meaning or discussion. Later, we'll consider situations where the full structure is important.

Whatever structure we depict as a selective representation, it is always still part of a *full structure* whose form is that of a fully connected graph.

Kinds of objects and structures

Structures and objects can be of any kind. A mathematical function is a structure. A concept is a thought structure. A thought is a structure. We can build new objects out of old structures and new structures out of old objects. Whatever apparently new structures we build, are constructed through the structure of the human brain. At the highest level of generality and abstraction, objects are thoughts of a particular kind.

There is a close relationship between mathematical structures that are structures of relations between numbers, and our conception of *objects* in general. That relationship persists in the relation between number-structures and the structure of our thought and thinking. This in turn is related to the structure of our brain, not just its physical structure, but its functional structures.

Morphisms

A morphism should not necessarily be thought of only as a mathematical function, but it also may sometimes be precisely that. It is a change to a structure or object according to a rule or set of rules. A *rule set* is itself a structure.

We can approach a morphism in terms of the relation or structure that it is, as $\xleftarrow{\;M\;}$ for example, or we can concentrate on the process of change and its direction, using a single-headed arrow, for example, $\xrightarrow{\;M\;}$. We use whatever is appropriate for the argument or discussion we are making.

A morphism $\mathbf{A}\xleftrightarrow{\;M\;}\mathbf{B}$ is the structure of relation of $\mathbf{A}$ to $\mathbf{B}$ that is *provided by* the morphism M, which is more structure. So in

$$\mathbf{A}\xleftrightarrow{\;M\;}\mathbf{B} \equiv \mathbf{A}\xleftrightarrow{\;R1\;}M\xleftrightarrow{\;R2\;}\mathbf{B}$$

the $\mathbf{A} \xleftrightarrow{R1} M$ is the relation of $\mathbf{A}$ with M and $\mathbf{B} \xleftrightarrow{R2} M$ is the relation of $\mathbf{B}$ with M.

In the case of a mathematical function, and where the morphism is reversible, then we may have

$$\mathbf{B} = f(\mathbf{A}), \mathbf{A} \xrightarrow{f} \mathbf{B}, \mathbf{B} \xrightarrow{f^{-1}} \mathbf{A}.$$

Then the following structures also apply:

$$\mathbf{A} \xleftrightarrow{R1} f \xleftrightarrow{R2} \mathbf{B},$$
$$\mathbf{A} \xleftrightarrow{R3} f^{-1} \xleftrightarrow{R4} \mathbf{B},$$
$$f \leftrightarrow f^{-1}$$

A morphism $\mathbf{A} \xleftrightarrow{M} \mathbf{B}$ may suggest something *happening*, as $\mathbf{A}$ changes to $\mathbf{B}$. In fact, it is a static arrangement. It is just a structure, and our idea of something *happening* arises from the idea of $\mathbf{A}$ changing into $\mathbf{B}$. It is in fact inseparable from the idea that $\mathbf{A}$ *changes into* $\mathbf{B}$. This is in turn intimately linked with our pre-existing ideas of *change* and *time*. In fact, $\mathbf{A}$ "changing into" $\mathbf{B}$ is just a relation of structures.

Our whole psychological idea of change as a process, is rooted in our experiential animal intelligence in which we experience time and change. This is not merely a matter of philosophy. We already know from physics, for example, that time itself is mathematically emergent from quantum correlations that are independent of time and space.

Our pre-existing ideas of *change* and *time* come from our experience, our experience of being, self, and world, provided by nature through the principle of the brain, which is a structure. Morphisms between structures are just structures.

A morphism that is not being considered as reversible may be written $\xleftrightarrow{\rightarrow}$ or $\xleftrightarrow{\leftarrow}$. To say, for example, $\mathbf{A} \xleftrightarrow{\rightarrow} \mathbf{B}$, does not however constitute a statement that the reverse morphism is not possible. It refers to a directional morphism in a particular structural description under consideration.

In the case of a non-invertible function such as $f_m(a,b) = a \bmod b$, where we have $f_m(a,b) = c$, then there is a structural morphism

$$\mathbf{A} \xleftrightarrow[\mathbf{F}]{\rightarrow} \mathbf{B},$$
$$\mathbf{A} \equiv (a \leftrightarrow b),$$
$$\mathbf{B} \equiv (c),$$
$$\mathbf{F} \equiv (f_m)$$

The morphism is a path from $\mathbf{A}$ to $\mathbf{B}$ *through the structure of the morphism.* The path passes through nodes in the structure, at which in the reverse direction, at least one bifurcation of the path is possible. For example, in $f_m(8,3) = 2$, there are many other possible values of a and b that through f_m will yield 2. The reverse path through the structure bifurcates into all these possibilities. This is because of the nature of the structure. So, for example, in the case of $f_m(8,3) = 2$, we have

$$(8 \leftrightarrow 3) \leftrightarrow M \leftrightarrow 2$$

At a superficial level it is easy to see that M is a set of rules applying to the left as its input, to produce the result on the right as its output. It is also easy to see why a reverse path from right to left cannot be encapsulated in a suitable "reverse set of rules", knowing, as we do, the rules of division and remainders, and being familiar with $\mathbb{R}$, the set of real numbers from which the numbers we are playing with, must be selected. However, in terms of structures, the one-way nature of the morphism is more complex, despite the ease with which we can conceive it.

Any function $f(\{x_n\})$ is a structure of relations (intervals) between the rules of f, which are a structure, and the structure of its domain $\{x_n\}$. For example, if the structure of f is $\mathbf{F}$, and the structure of $\{x_n\}$ is $\mathbf{R}$, the structure of the real numbers $\mathbb{R}$, then the meaning and operation of $f(\{x_n\})$ lies in the structure $\mathbf{F} \leftrightarrow \mathbf{R}$. Furthermore, $\mathbf{F}$ contains rules involving mathematical operations, which are still only partially understood, and **R** is the structure of a set $\mathbb{R}$ of objects called real numbers whose rules of structural relations (intervals) haven't even been fully worked out.

The structure $\mathbf{F} \leftrightarrow \mathbf{R}$ is already a complex (in the sense of complexity) structure, so $\mathbf{A} \underset{\mathbf{F}}{\overset{\rightarrow}{\longleftrightarrow}} \mathbf{B}$ is a complex structure, and this morphism is a path through a complex structure.

Symmetries and morphism of structures

Symmetries, and morphisms of structures, may be regarded as different structures. Symmetry morphism is in the form:

$$\mathbf{A}_1 \xleftrightarrow{SM} \mathbf{A}_2 \equiv \left(\mathbf{A}_1 \xleftrightarrow{R1} \mathbf{O} \xleftrightarrow{R2} \mathbf{A}_2 \right) : \mathbf{O} \equiv \mathbf{A}_1 \cap \mathbf{A}_2$$

where **O** consists of at least one object in both $\mathbf{A}_1$ and $\mathbf{A}_2$. The same symmetry principle can apply to spaces, e.g. :

$$\left(\mathbf{A} \leftrightarrow \mathbf{S}_1 \xleftrightarrow{R1} \mathbf{O} \xleftrightarrow{R2} \mathbf{S}_2 \leftrightarrow \mathbf{B} \right) : \mathbf{O} \equiv \mathbf{S}_1 \cap \mathbf{S}_2 .$$

Notations

A structure of many objects may be written in short by listing its objects in parentheses following a vertical arrow, for example:

$$\mathbf{S} \equiv \updownarrow \left(\mathbf{A}, \mathbf{B}, \mathbf{C}, x, y, z \right) .$$

Alternatively, the objects may be numbered, and the structure represented in the form $\updownarrow \left(\{ \mathbf{O}_n \} \right)$, or just $\updownarrow \left(\{ \mathbf{O} \} \right)$. The vertical arrow is a reminder that we are dealing with a structured set with intervals, as a fully connected graph, whose structure is in terms of structure theory. These notations can be used when we want to refer to the importance of the full structure. For example, we might write $\mathbf{S} \equiv \updownarrow \left(\mathbf{A}, \mathbf{B}, \mathbf{C}, \mathbf{D} \right)$ for a structure of 4 objects, where all 6 intervals (relations) are important to the discussion. Otherwise, we might just say

$$\mathbf{S} \equiv \mathbf{A} \leftrightarrow \mathbf{B} \leftrightarrow \mathbf{C} \leftrightarrow \mathbf{D}$$

if we are talking about specific intervals. In this case the full structure is still implied, but not so relevant to the discussion.

Parentheses may be used to enclose structure representations, which in some circumstances can be in associative and commutative ways, for example:

$$\mathbf{A} \leftrightarrow \mathbf{B} \leftrightarrow \mathbf{C} \equiv \big[(\mathbf{A} \leftrightarrow \mathbf{B}) \leftrightarrow \mathbf{C} \big] \equiv \big[\mathbf{A} \leftrightarrow (\mathbf{B} \leftrightarrow \mathbf{C}) \big] \equiv \big[(\mathbf{A} \leftrightarrow \mathbf{C}) \leftrightarrow \mathbf{B} \big].$$

Structures can also be regarded as distributive in the following way:

$$\big[(\mathbf{A} \leftrightarrow \mathbf{B}) \leftrightarrow \mathbf{C} \big] \equiv \big[(\mathbf{A} \leftrightarrow \mathbf{B}) \leftrightarrow (\mathbf{A} \leftrightarrow \mathbf{C}) \big]$$

Naturally, this doesn't apply where intervals are directional morphisms.

However, structure conceptions do not translate directly into consistent structures of symbols and rules for their manipulation. For this reason, structures, at least at this stage of their depiction, are not an algebra in the ordinary sense.

Gadgets and devices

Structures may be manipulated by structural operators and gadgets that are comparable (but different) to other mathematical operations. A *structural function* looks superficially like a conventional mathematical function but manipulates *structures* directly. A *gadget* is a construct composed of structural functions, and a *device* may be composed of many gadgets.

Despite this terminology, we can view the complex *natural mechanisms* of both non organic physical systems and life forms, as composed of *natural gadgets and devices.* In general, evolution can itself be viewed as a natural device.

If we have the directional morphism $\mathbf{S}_1 \xleftrightarrow{\;\rightarrow\;} \mathbf{S}_2$, we may also write it as $f(\mathbf{S}_1) \xleftrightarrow{\;\leftrightarrow\;} \mathbf{S}_2$ or $f(\mathbf{S}_1) \equiv \mathbf{S}_2$. Here, f is a *structural function*. The $\xleftrightarrow{\;\leftrightarrow\;}$ arrow or the $\equiv$ sign is the structural equivalent of the = sign. It means specifically that the objects either end of it are one and the same single object or structure, written out twice (two instantiations).

We cannot, in general, simply say $\mathbf{S}_1 = \mathbf{S}_2$, without further gadgets, because in structures we must be clear about whether we are talking about two instances of the same object (distinct identical objects) in a structure, or two written representations of the same single object or structure. If $\mathbf{S}_1$ and $\mathbf{S}_2$ are one and the same object then we would have either $\mathbf{S}_1 \equiv \mathbf{S}_2$ or $\mathbf{S}_1 \xleftrightarrow{\;\leftrightarrow\;} \mathbf{S}_2$. If there are two distinct, identical instances of $\mathbf{S}_1$, then we would write $\mathbf{S}_1 \leftrightarrow \mathbf{S}_1$.

Whilst we cannot use the = sign to specify that two structures are the same structure or two instances of the same structure, we can use it to show equality between two gadgets, meaning that they are equivalent and interchangeable.

A more complicated gadget is a *device* that we can specify in terms of *rules*, and treat it as a structural function that acts on one or more structures. Essentially, a device can be thought of as like an *object oriented program*, and the structures on which it operates, as objects within the program.

A *structural iterated function system* (SIFS) is a powerful device both for manipulating structures and for exploring iterated structure creation and behaviour. Like the IFS of the kind that might be used to generate fractals, it can be thought of as a program that iteratively applies rules of change to a mathematical structure or object, taking its input from its output. However, unlike a computer program, the SIFS can, like an IFS, be *infinite* in its iterations. An infinite SIFS can be considered as an object in its own right that can be "equal to" the structure that arises from its infinite iteration.

Reference to sets - objects versus intervals

Although the relations, intervals, or graph edges, in a structure, are objects in their own right, if we refer to a *structure* as a *set*, or talk in terms of sets about the objects from which a structure is composed, or if we refer to a structure's *objects*, then we refer only to the set of nodes, and not the set of relations between nodes.

An object that has no intervals attached to it and no defined further structure, is an *abstract* and *uninstantiated* object. Every object represented in a structure is an *instantiation* of an object.

Spaces and identical objects

Two *identical* objects in a structure, unless shown to be related by the $\equiv$ sign (or a $\overset{\leftrightarrow}{\longleftrightarrow}$ relation) are not the *same* object, but two separate *instantiations* of the object.

Two *identical* objects can only be *adjacently* instantiated in the same structure if there is a *space interval* between them. The interval may turn out to be some structure or object that we do not wish to regard as a space in some other sense, but unless so specified, we regard it as a *space interval* in the structure.

A single object with no specified further structure is *uninstantiated*. In order to instantiate it, it must be specified at least in relation to a space object. So in the "protocol" of structure descriptions an object **A** with no further specification is uninstantiated. The least we must do to instantiate it is to specify it together with a space object $\mathbf{S}_P$, such as $\mathbf{A} \leftrightarrow \mathbf{S}_P$.

Essentially, this is the same as writing $\mathbf{A} \leftrightarrow$, but the nature of a *relation* $\leftrightarrow$ depends on objects, in order for the concept of a *relation* to make sense. So although any object or structure has implicate in its instantiation further intervals to unspecified objects or structures, in a way somewhat similar to how we might say "all systems are open", *writing* an arrow *without* specified objects adjacent on it, isn't appropriate in structure representation.

Structural spaces - some consequences

In a structure $\mathbf{A} \leftrightarrow \mathbf{A}$ of two identical objects, the interval $\leftrightarrow$ is by definition a *space*, unless otherwise specified. A space is a structure, and if we wish to specify this interval further, then we may not necessarily wish continue to regard it as a space. We may also write $\mathbf{A} \leftrightarrow \mathbf{O} \leftrightarrow \mathbf{A}$, where $\mathbf{O}$ is an object or a space, depending on what we want to specify.

If in $\mathbf{A} \leftrightarrow \mathbf{O} \leftrightarrow \mathbf{A}$ we specify $\mathbf{O}$ is a *space*, then because the $\mathbf{A}$s are two identical instantiations the same object, the two intervals (the relation to the space) *must differ*. However, they need not differ if we specify $\mathbf{O}$ as another object not identical to $\mathbf{A}$.

In the structure $\mathbf{A} \leftrightarrow \mathbf{A} \leftrightarrow \mathbf{A}$, the intervals are spaces. They can be two identical instances of the same space, but this doesn't mean the relations of each instance of the object $\mathbf{A}$ has the same relation to the space. In fact they cannot. The intervals specify that the $\mathbf{A}$s are related to each other by the same space. There is here no further specification. If we now specify further, we can write out the spaces as space objects $\mathbf{S}$, in the manner that we illustrated earlier:

$$\mathbf{A} \xleftrightarrow{\ \alpha\ } \mathbf{S} \xleftrightarrow{\ \beta\ } \mathbf{A} \xleftrightarrow{\ \chi\ } \mathbf{S} \xleftrightarrow{\ \delta\ } \mathbf{A}$$

where now the intervals are different. However, we could also have

$$\mathbf{A} \xleftrightarrow{\ \alpha\ } \mathbf{S} \xleftrightarrow{\ \beta\ } \mathbf{A} \xleftrightarrow{\ \alpha\ } \mathbf{S} \xleftrightarrow{\ \beta\ } \mathbf{A} .$$

In the structure

$$\mathbf{A} \leftrightarrow \mathbf{O}_1 \leftrightarrow \mathbf{B} \leftrightarrow \mathbf{O}_2$$

$\mathbf{O}_1$ or $\mathbf{B}$ may be regarded as spaces, according to need. $\mathbf{A}$ or $\mathbf{O}_2$ may be spaces if we consider the full structure. We can say

$$\mathbf{A} \leftrightarrow \mathbf{O}_1 \leftrightarrow \mathbf{B} \leftrightarrow \mathbf{O}_2$$

or

$$\left(\mathbf{A} \leftrightarrow \mathbf{O}_1 \leftrightarrow \mathbf{B} \right) \leftrightarrow \mathbf{O}_2$$

or

$$A \leftrightarrow \left(O_1 \leftrightarrow B \leftrightarrow O_2\right)$$

and so on, as necessary, but in each case the meaning of the structure in terms of which objects we may or may not regard as a space, is likely to be different.

Superposition

The superposition of structures can be written as:

$$\overline{S_1 S_2} \equiv S_3$$

where the bar over the LHS indicates the objects placed side by side under it, are superposed.

Relation of structures to natural objects

An object is regarded as a structure, despite that it may be specified to consist of only one object. An object is a structure also regardless that it may be defined as "structureless" in some other context, such as physics. Symbolically, a structure S is a set of objects $\{O_a\} : O_a \in S$, and O_a may be singular, *i.e.* $O_a \equiv S$.

Algebraic structures

If we write the polynomial $y = x^2 + 2x + c$, then in the *written* structure each x is a distinct instantiation of the written object x, but it is only a distinct instantiation of the written symbol. We can say that this polynomial, as written, is a structure

$$y \leftrightarrow = \leftrightarrow \left(x^2\right) \leftrightarrow + \leftrightarrow (2x) \leftrightarrow + \leftrightarrow c,$$
$$\left(x^2\right) \equiv \left(x \leftrightarrow X \leftrightarrow x\right),$$
$$(2x) \equiv \left(2 \leftrightarrow X \leftrightarrow x\right)$$

which is one way of writing the structure of the written equation. But in the equation structure the x variables all represent one and the *same* mathematical object, rather than each appearance representing a distinct instantiation of an identical object in a structure.

This is a different *modus operandi* to native structure representations. So if we want to make sense of the *meaning* of a written equation as a structure, as distinct from its *written* structure, and in the sense of structure theory, then we have to account for the appearances of the variables and constants in this way.

The mathematical object that the equation represents isn't the equation as written. It is the structure of relations between 4 numbers. We could even write in short the whole structure of the mathematical objects to which the equation refers as $\updownarrow\left((y,x,2,c) \leftrightarrow (+,\wedge,X,=)\right)$. The full structure contains intervals between the nodal objects $y,x,2,c$, and the intervals specify the relations $+,\wedge,x,=$, that the equation represents.

What an ordinary written mathematical function allows us to do, is to represent in a coded written structure called an equation, a pure *mathematical form* that consists of rules of relation between objects, where the objects are usually the stuff of numbers and number ranges.

All written mathematical equations are codes for mathematical structures, where the written structure is of the kind $\updownarrow\left(\{\mathbf{O}_n\}\right)$, in which there is a set of n related distinct written objects $\mathbf{O}_n$, that are numbers, or placeholders for numbers called variables. The written equations are not the mathematical structures themselves, but coded representations, not least because many of the objects $\mathbf{O}_n$ are only distinct in the written representation. The written representation is an artefact of the way our intelligence, evolved through nature, perceives and comprehends the phenomenon of the mathematical structure itself.

Written mathematical structures are mapped *selective representations* of natural mathematical structures, mapped through the conventions, into the code with its symbols and its rules. They involve the functioning of our own brain and intelligence. Functions, rings, groups, categories, functors and natural transformations, as written, are all mappings of the corresponding natural mathematical structures and pure forms, rather than being those structures and pure forms themselves.

Even natural transformations are forms of structure, beyond the appearance of numbers themselves, that are *selective representations* of structures that exist in our own intelligence. The functioning of our own brain and intelligence cannot be isolated from what we comprehend, and what we comprehend cannot be isolated from our natural intelligence. Our comprehension of written mathematical objects and structures is a selective and biased comprehension of the operative nature of our own intelligence.

Algebraic structures are networks of algebraic objects that are either numbers, placeholders for numbers, or structural gadgets and devices for applying structural morphisms to the structure of numbers. Algebraic structures are not part of nature, or how nature works, isolated from our intelligence. They are only objective, as part of our intelligence understood as an object.

Meaning

A structure as we depict it has associated with it *meaning*, which is the comprehension and understanding we have of it. Equations are structures. The equation $w = f(x,y,z)$ can be represented

$$w \leftrightarrow = \leftrightarrow \left(f \leftrightarrow \left(x \leftrightarrow \left(y \leftrightarrow z \right) \right) \right)$$

where the x,y,z in this representation are commutative. What we have written is a *selective extract* from the full native mathematical structure in the form of a fully connected graph, that the *written structure* selectively represents.

The whole mathematical structure contains the six objects, $w, =, f, x, y, z$. If we have more information about f then we may write the extract differently. For example, if we know that $w = x^2 + yx + \log_e z$ then we can write

$$w \leftrightarrow = \leftrightarrow \big((x \leftrightarrow 2) \leftrightarrow + \leftrightarrow (y \leftrightarrow x) \leftrightarrow + \leftrightarrow (e \leftrightarrow z)\big)$$

as our extract. Now we also have more information about the intervals. For example, in the last parentheses we know that the interval means e and z are related as $\log_e z$. We can even write the arrow if we want to, with a specifier to stand for the information on how the objects are related, such as

$$e \xleftrightarrow{\ Log(z)\ } z \, .$$

Each object in a structure has a meaning in the context of the meaning of the structure as a whole, whilst the intervals are placeholders for the information on how those meanings of the objects are related.

The structures of *written* mathematics may *translate* into native mathematical structures, but are not the native mathematical structures themselves. They translate in part through the structure of our own intelligence. They tend to force us to think in certain ways, based on the written structures. This is why very often, the concept of infinity, which as a concept is an object, is often thought of as though it were a number, or algebraically or arithmetically manipulable, when in fact, as we already know, it is not.

A good deal of mathematics consists of establishing natural rules of relations between objects without necessarily fully understanding the objects or what they mean. For example, what numbers of a given type are, or what they mean, as uninstantiated objects, does not necessarily need to be fully understood or universally agreed upon, before effective mathematics, as a tool, is possible.

Mathematics is about the rules of mathematical structures, and the rules of morphisms between them. Abstract algebra is already about mathematical structures, but they are not structures seen in the way structure theory sees structures. The structures of abstract algebra themselves stand to be deconstructed by structure theory.

We often introduce objects whose meaning is understood in the context of the structure, without necessarily having to know everything about the objects and their structures. The deeper nature of these objects is not necessarily something we need to fully understand, in order to use them in structures. This can be easily seen in science in concepts such as dark matter and dark energy introduced into the structures of cosmology. Such introductions of new objects into the structure of our comprehension, before understanding the objects fully, is a repeated feature through the history of science.

The same basic principles of the usefulness of objects apply to other structures, such as the structure of the atom, the structure of molecules, the structure of cells, and so forth. We can attain a good knowledge of chemistry by instantiating concepts like atom, charge, proton, neutron, electron, orbit and shells, without necessarily having any deeper knowledge of any of these component objects that can be instantiated in the structure of the atom. Historically, in the development of science, this is precisely what happened. The progress of scientific knowledge of nature has been partly a process of increasing *definition*.

Personal and group belief systems, and personal comprehension systems are also structures. They rely on a set of concept objects that are related, and are instantiated into a person's belief structure, or comprehension of something, without necessarily fully understanding, or perhaps even properly understanding at all, the concept-objects that are instantiated. Even the individual relations between the objects may not be really understood. But in a structure, if there is *sufficient* meaning - whether it is true or false - then the structure is effective.

Human psychology itself is a structure. As we come to greater knowledge of the human brain, we will come to greater knowledge of it as a structure, because that's what the human brain is.

Internal and external

The notions of "internal" and "external" relative to objects, are not concepts that are native to structures. A *ball* may be defined as an object, $O \equiv ball$. There is a certain amount of meaning already inherent in this object, when named. Psychologists call this meaning the *apperception mass* associated with the word *ball*. The momentum and position p, x of the ball are then further objects, and also *furnishings* of O . In this case, $ball, p, x$ are objects in a structure. If we say the ball has *greenness*, then *green* or *greenness* are further objects that are also furnishings of O .

It may be convenient and/or intuitive to consider certain furnishings as "internal" or "external" to O. We might say, for example, that the ball has a distance d from another ball, O_a, and that $d \not\subset O$. We might do this because we don't want to regard *d* as a property of the first ball, O . When we are considering structures this question of "internal" versus "external" is one of sets, and has its place, but it is like a "translation" from the native "language" of structures to the language of objects understood according to sets.

Identity, objects and spaces

As we have seen, in structure representations the concept of two identical objects that are distinct, meaning they are not one and the the same object, has no meaning on its own, without further specification. Distinction requires another object, to which the two identical objects must have different relations. That other object may be a space.

Our experience of our world places everything *in* a space of some kind, a space in which things are separated, whether that be physical space, geometrical space, or in many mathematical descriptions some kind of system space. In terms purely of objects and relations, a space in the simplest terms is an object. As something occurring in the description of a structure or system in natural phenomena, or in the mathematical underpinning of such descriptions, a space is a structure.

In structure theory two distinct but identical objects can only exist in relation to a space or another object, and a single object can only be *instantiated* in its first instance, together with a space. We don't have to begin with a duality of non space objects, but we may. If we begin with one object **A** , then it must be by specifying **A** ↔ **S**, where the second object is a space. A space is just a conceptual name for an object or structure, which in mathematical descriptions fulfils certain requirements in relation to other objects and structures.

There is of course an applied counterpart to this. In modern cosmology stars and planets are regarded as objects, and *spacetime* as the space *in which* such objects occur. But black holes, which are part of the structure of spacetime, are also spoken of and conceived as objects. Einstein's equation shows both spacetime and these objects "within it" to be part of the same mathematical structure.

Although we can use our mathematical descriptions to model natural phenomena effectively, there is very little recognition of the interchangeability of objects and relations in natural phenomena, that is, outside mathematical calculations themselves, until we come to quantum mechanics. And even then, the intuitive notion of objects and spaces, despite the mathematical structures, is still clung to at least for the prosaic reason that particle detection is about the detection of particles as objects in the space and time of the everyday world.

If we define a set $\{\mathbf{O}_n\}$ as a space, as in often done in mathematics, then in the view of structure theory there is a space (a set of structure theory spaces) within that space. The continuity of points $\mathbf{O}_n$ as a continuum, that in conventional mathematical theory comes down to the issue of a continuity of number intervals, becomes in structure theory (as we shall be seeing), a matter of how structures, and numbers themselves, as structures, are constructed.

A simple intuitive illustration

Consider a collection of theoretical, truly identical balls on a table, in a structure that consists only of balls and table. If they are truly identical, the only way in which they can be separately identified in this structure, is through the fact that they occupy different positions on the space of the table. The space of the table is a canonical form of the more generalised concept of a mathematical or number-space, as a structure relative to which other structures or objects appear. In structure theory, it is not that the balls appear *in* the space, but rather, that they have *relations* to it.

If an object such as a theoretical ball is *fully defined* and specified by a set of properties, then to repeat the specification with no further specification is not to specify a duplicate ball, but to specify the same ball twice. In order to instantiate two identical balls we must have some other property or space relative to which each instantiation of the ball specification has a different relation. This could be for example, colour, mass, or position.

Colour and mass can also constitute a space. In a theoretical structure a number of balls could occupy the same position on the table, something that rules of material balls on material tables prohibit. If they are identical balls, then without further specification they are structurally one and the same object. The multiplicity is one of merely theoretical instantiation. If they differ in colour, then they are distinct in colour-space, even though they occupy the same position on the table.

The distinctness of objects in general, depends on a space. The distinctness of *identical objects* depends either on a space that provides the difference between the objects as furnishings to the objects, or is *structure space*.

Structure and category

Commonly, the additional *furnishing* or space is presumed or implied. Until such structure is *specified*, it cannot influence the status and meaning of a declared structure. In structure theory two objects **A** and **A** refer to the same one object, just as they would as two instances of the same variable in an equation. In contrast, two instantiations of an object as two identical objects must have a relation to each other, as $\mathbf{A} \leftrightarrow \mathbf{A}$. By default, this is a space.

So in *structures* dualities of objects that are otherwise identical are multiple *instantiations* of objects, each necessarily with a unique relation to a space. Unique relation to a space is the only means by which a duality of otherwise identical objects can occur in a structure. A space is a structure that may or not not be considered to contain as part of its own structure *properties of the objects* that are related it to it.

If two identical balls $\mathbf{B}_1$ and $\mathbf{B}_2$ are identically specified, without specifying some further space to which they have unique relations, then both specifications are for the same ball. If, now, the balls are assigned different colours, then the colours can be regarded as a space.

As sets, we would say $\sim(\mathbf{B}_1 \cap \mathbf{B}_2)$ is the colour space. A structured sets, we would say $\mathbf{B}_1 \cup \mathbf{B}_2$ is the structure. As structures, we can say $\mathbf{B}_1 \leftrightarrow \mathbf{C} \leftrightarrow \mathbf{B}_2$, where $\mathbf{C}$ is the colour space.

If we were to say

$$\mathbf{B}_1 \leftrightarrow \mathbf{C} \leftrightarrow \mathbf{B}_2 \leftrightarrow \mathbf{C} \leftrightarrow \mathbf{B}_1$$

then we have merely overused the notation. But if we say

$$\mathbf{B}_1 \leftrightarrow \mathbf{C} \leftrightarrow \mathbf{B}_2 \leftrightarrow \mathbf{C} \leftrightarrow \mathbf{B}_3 \leftrightarrow \mathbf{C} \leftrightarrow \mathbf{B}_1$$

then we have stated something different that we could also write in a more condensed way as

$$\mathbf{B}_1 \leftrightarrow \mathbf{B}_2 \leftrightarrow \mathbf{B}_3 \leftrightarrow \mathbf{B}_1$$

where now if every $\mathbf{B}_n$ is identical, then the intervals must be spaces (in this example, a colour space). We have specified the full structure, and the structure has the requisite minimum of three objects necessary to be a "closed form", or *cycle* of objects and intervals. We could just illustrate a triangle of the objects.

The intervals here specify that the relation of each pair of objects adjacent on an interval is one whereby each object has a unique relation to a space, and that it is the same space for every pair. So if we now re-write each $\mathbf{B}_n$ combined together with its relation to the space, as a single new object $\mathbf{O}_n$, then we can swap to category theory notation and we would have the relations:

$$\mathbf{O}_1 \xrightarrow{\text{M1}} \mathbf{O}_2 \xrightarrow{\text{M2}} \mathbf{O}_3,$$
$$\mathbf{O}_1 \xrightarrow{\text{M3}} \mathbf{O}_3$$

Each object is identified by its relation to the space, which is a relation or morphism through which it retains the same identity. M1, M2, and M3 are morphisms or relations through the colour space. We also have that $M2 \circ M1 = M3$, and so the structure commutes. The structure is a category, albeit trivial. But what we can also see, is that the category is a structure of objects and spaces, in the structure theory sense.

Structures and intelligence

Because valid mathematical structure representations as far as we know them, must have proper *meaning*, we can say that a valid mathematical structure, corresponding to a written representation, only exists as a mathematical structure if its written representation has this required meaning. Thus, for example, the written structures:

$$a = 2, b = 2, a + b = 5$$

$$\text{and}$$

$$\forall \left(< \notin \pi \right)$$

lack the required meaning and we can regard them as non-existent, as actual mathematical structures, as distinct from the written expressions. However, they might become a representation of an existing structure if we can assign suitable meaning to them.

The meaning of any structure, in as much as we can represent it, requires our own intelligence. A structure $\updownarrow \left(\left\{ \mathbf{O}_n \right\} \right)$ is not in itself the sole repository of the *meaning* we may find in the structure, because this meaning pertains to our intelligence. This is perhaps the most overlooked aspect of all of mathematics.

Being a graph, a mathematical structure $\updownarrow \left(\left\{ \mathbf{O}_n \right\} \right)$ may include subgraphs or *substructures*, and being a structure, it has potential intervals to other structures. In itself, a structure $\updownarrow \left(\left\{ \mathbf{O}_n \right\} \right)$ of the kind we normally encounter does not specify its own existence, even if it has proper meaning to us. Rather, we must propose a structure $\updownarrow \left(\left\{ \mathbf{S}_q \right\} \right)$ with a degree of complexity appropriate to our own intelligence, as the human brain, such that

$$\updownarrow\left(\left\{\mathbf{O}_{n}\right\}\right)\leftrightarrow\updownarrow\left(\left\{\mathbf{S}_{q}\right\}\right)$$

fulfils this requirement. This further structure must, if it accounts for our own comprehension of the meaning, in some way account for the nature of our own intelligence.

A structure $\updownarrow\left(\left\{\mathbf{O}_{n}\right\}\right)$ should not therefore be regarded as being in itself a pure *mathematical form*. In itself it consists of objects and intervals that first require intelligent conception. The conception can only provided by the extended structure

$$\updownarrow\left(\left\{\mathbf{O}_{n}\right\}\right)\leftrightarrow\updownarrow\left(\left\{\mathbf{S}_{q}\right\}\right)\equiv\Omega$$

which itself is the structure of both our conception of this extended structure, and the conception of mathematical structures of the kind $\updownarrow\left(\left\{\mathbf{O}_{n}\right\}\right)$ in general. This object Ω is a *limitation relation* on our comprehension of mathematical form. The pure mathematical form of which a structure $\updownarrow\left(\left\{\mathbf{O}_{n}\right\}\right)$ is an *expression* is only *partially* represented by $\updownarrow\left(\left\{\mathbf{O}_{n}\right\}\right)$, and requires Ω for its full representation.

Even the full representation Ω cannot be regarded as the pure mathematical form, because mathematical structures are in general structures of relation between objects whose "substance" is the pure substance of *Number*. In other words pure mathematical form is form that takes place in the substance of Number. We must first understand this pure substance of Number as the basis of mathematical structure. To get to see pure mathematical form in relation to our intelligence, and *vice-versa*, we cannot merely define the pure substance of Number in terms of more mathematical objects (like, for example, sets). We have to go beyond mathematical objects.

So even the structure of Ω rests on the pre-existing substance of Number. We can find *structure* in Number, but still this structure is in the form of $\updownarrow\left(\left\{\mathbf{O}_{n}\right\}\right)$, and even as Ω it is still secondary to the substance of number. It is the substance of Number that enables Ω, both by enabling the $\updownarrow\left(\left\{\mathbf{O}_{n}\right\}\right)$ part of it that exists as the object structures we can comprehend, and by enabling the $\updownarrow\left(\left\{\mathbf{S}_q\right\}\right)$ part of it that is the arising of our own brain and intelligence in which this comprehension takes place. We should regard this as fundamental to nature, from a scientific point of view.

The substance of Number is what enables the principle of *duality*, whereby different or identical objects can exist, that are distinct. The substance of Number is also what enables the principle of a *space*, whereby identical objects can be distinguishable as different objects. So the "pure form" or substance of Number, if we can put it like that, rather than further structures, even rather than the structures to be found in pure Number, is what holds the key to all structures, rather than *vice versa*.

Natural laws

If we talk about a *coffee cup*, we know what we mean. Even before we have furnished the object we have just named, with more properties. But we only know what we mean, because the object is already connected, in our mind, to more objects. In psychology this network of objects is called an *apperception mass*. In structure theory, it is a *structure*.

Our comprehension of the object named *coffee cup* only happens because there is something more than just connections in the structure. The connections are alive with rules and operations. As are the connections in the network of our brain. In structure theory these connections in structures are the *relations* or *intervals* between objects.

Consider the case of electrons. The electrons are usually considered to be identical, except for additional furnishings that may be different for different instances of an electron. These furnishings are quantum properties. They are part of the mathematical *space* - as defined in structure theory - to which the basic concept-object *electron* has relations. But mathematically, the electron *is* the whole structure of all its properties.

In the case of entangled quantum particles, two "separate" particles are in fact one mathematical structure. The puzzlement that often arises over the natural behaviour of quantum entanglement is only due to the attempt to understand the deepest nature of our world in terms of naive realism's "objectivity", rather than as structures of objects, where objects are objects in the sense of structure theory.

In structure theory the separation of entangled quantum particles is part of the structure of the space to which they have relations, and their now proven ability to synchronise states independently of *cause and effect* in metric space, is a feature of the *overall* structure in which their behaviour arises.

If we are scientifically describing the behaviour of natural phenomena, then we will always be using *object structures*, of the kind structure theory describes, whether or not we recognise them as such. The workings of nature ensures that the very process of our dealing with structures in the endeavour to understand the workings of nature, will of course draw us towards what we might loosely but usefully call "attractors" in the combination of these structures and our thinking about them. These are what we come to know at any time in the development of science, as our understanding of nature's laws.

Intelligence and the brain

The furnishings of an object, any properties assigned to it, are themselves objects that have *relations*. In the example of the ball, the momentum and position p,x, of the ball, having been assigned to the ball, are objects that are part of the *structure* in which the object *ball* appears. That structure may be a *space*. We see instances of this in concepts such as *parameter space*, *system space*, *phase space*, and *Hilbert space* as used in quantum mechanics.

A *structure* **S** can include the furnishings of its objects, the relations between its furnishings, and the relations of its objects to the furnishings. Where any of those furnishings are relations between objects in **S** to any set $\{P_n\}$ of objects where $\{P_n\} \supset \mathbf{S}$, then **S** must also be a substructure of another structure $\mathbf{L} : \{P_n\} \subseteq \mathbf{L}$.

This isn't something confined to a hypothetical collection of balls. In fact, *all* structures **S** that we study as other than, and separate from, our intelligence itself, are structures subject to just such a condition. In our objective view of nature any such structure **S** in nature consists of objects that *do* have relations to other objects $\{P_n\}:\{P_n\} \supset \mathbf{S}$, where $\{P_n\}$ is the set of objects that contains the structure of our own intelligence, *in which the understanding happens.*

The very *understanding in our intelligence,* of an object or structure as an object that is *other than our intelligence,* itself requires *relations* between that structure $\mathbf{S}$ and the structure $\mathbf{I}$ of our own intelligence and brain activity.

Although it may often be assumed that $\mathbf{S} \cap \mathbf{I} \neq \varnothing$, structure theory implicitly recognises that our own intelligence is *inevitably* already part of the objective structure we may be studying, and that any structure under consideration, is part of the same structure that gives rise to our own intelligence.

Now this may *look* initially like a matter of philosophy rather than mathematical structure, and as if it is something divorced from hard science. But as it happens, it is not just a quirk of structure theory, generated merely by its own rules or axioms. There is already a very ordinary way in which this is already expressed, in non mathematical terms, which is simply to state that *Cartesian duality is incorrect.*

The fact is, that as far as mainstream modern neuroscience is concerned, Cartesian duality is indeed incorrect, and not a modern scientific view. Furthermore, Cartesian duality, the duality of the observer and the observed, is not consistent with quantum mechanics, our most successful scientific theory.

So this situation is far from being a merely philosophical proposition, or theoretical construct, or artefact merely created by structure theory. It is already present in actuality, in hard science, in the *principle of the brain* as understood in modern neuroscience, as the relation between the brain as a system, our intelligence, and the world we experience, and in which we live. It represents the true, scientifically demonstrable relationship, between the objects of both the world and our mathematical understanding, and what modern neuroscience calls the "internal model" that our brain creates.

Constructions

General principles

The interval arrow spans between two objects (structures) that are not one and the same object, but may be distinct, identical instantiations of it. If we put a letter such as *M* on the arrow, it can further specify a relation, or and/or indicate a morphism between the objects. If we add another directional arrow, it can indicate the direction of a morphism, or also perhaps that the morphism is not considered reversible.

An arithmetical operation of addition on two numbers a, b, can be written $a \leftrightarrow + \leftrightarrow b$, where $a, b, +$, are all objects (they are representations of objects), and the arrows are intervals. The equation $a + b = c$ has the written structure $(a \leftrightarrow + \leftrightarrow b) \leftrightarrow = \leftrightarrow c$. Note that the *equals* sign is an object. The *concept of mathematical equality* is an object. Any written symbol is a symbol for an object. Every object is a concept, and every concept is an object.

The arithmetical convention of *associating* a *sign* of plus or minus with a variable or constant is just the use of one structure with two objects, where the sign and the symbol placeholder for a number, are the two objects.

So for example

$$-x \equiv (- \leftrightarrow x)$$
$$+x \equiv (+ \leftrightarrow x)$$

where the LHS is the usual conventional expression, and the RHS is the structural representation. What we cannot do is is use the *equals* sign in the conventional way, in such a mixed representation. The *equals* sign can *only* be represented as an object in a structure, and the $\equiv$ sign cannot, by this convention, be part of a structure. In mixed representations "$\equiv$" means "is defined as the structure", or "are one and the same instantiation". It does not mean "is identical to" with the normal mathematical meaning.

So in the structural representation $(a \leftrightarrow + \leftrightarrow b) \leftrightarrow = \leftrightarrow c$ it can simply be taken as given that, for example, $a \equiv (+ \leftrightarrow a)$, just as in an ordinary equation. The parentheses indicate a structure considered as a single object. We can use parentheses to contain objects that must be grouped together as a structure, because as the whole structure it has a relation to one or more other objects or structures. We can easily refer to the structure of a mathematical expression or equation without specifying the structure in any detail, by enclosing in parentheses, such as $(a = b)$ or $(y = f(x))$.

As a structure, the equation $1 + 1 = 2$ is

$$1 \leftrightarrow + \leftrightarrow 1 \leftrightarrow = \leftrightarrow 2$$

The intervals are the "differences" between the objects, but they are not by definition the arithmetical differences. The arithmetical difference is itself a number, and itself an object. The arrow represents an object too, but that object is the *relation* between the two objects it separates. One possible aspect of this relation is the arithmetical difference, but that is only one possible aspect of it.

Relation is a feature of two objects that is itself an object that is *instantiated* or describable in some cases, as arithmetical difference. But even in the case of two *numbers*, their arithmetical difference doesn't encapsulate the structural interval between them.

Equation structure

The ordinary functional equation $y = f(x)$ used as an expression of morphism, has the same meaning as the morphism notation $x \xrightarrow{f} y$. Structurally, $(y = f(x)) \equiv (x \underset{\rightarrow}{\xleftarrow{f}} y)$.

A function or morphism is an object that takes an input and gives an output, by embodying a structure of relations between mathematical objects, into which the input in included. The morphism structure $x \xleftrightarrow{f} y$ is

$x \overset{\alpha}{\longleftrightarrow} f \overset{\beta}{\longleftrightarrow} y$ where f is that structure, α is the relation specifying the x as the input, and β specifies that the relation between y and f is that y is the structure of the output of f. The most important thing here is that *a morphism is a structure.*

So $x \overset{\alpha}{\longleftrightarrow} f$ is the structure of $f(x)$, and

$$x \overset{f}{\longrightarrow} y \equiv \left(x \overset{f}{\underset{\rightarrow}{\longleftrightarrow}} \left(x \overset{\alpha}{\longleftrightarrow} f \right) \right)$$

In the structure $\mathbf{A} \leftrightarrow \mathbf{=} \leftrightarrow \mathbf{B}$ the $\mathbf{=}$ object is a structure that identifies two instances of the same number object $\boldsymbol{n}$, one apparently as a substructure in each of the objects $\mathbf{A}$ and $\mathbf{B}$. Essentially, the $=$ sign is an interval $\overset{=}{\longleftrightarrow}$ in the *equation structure*

$$\left(\mathbf{A} \leftrightarrow \boldsymbol{n} \overset{=}{\longleftrightarrow} \boldsymbol{n} \leftrightarrow \mathbf{B} \right),$$

$$\mathbf{A} \leftrightarrow \left(\boldsymbol{n} \overset{=}{\longleftrightarrow} \boldsymbol{n} \right) \leftrightarrow \mathbf{B}$$

where

$$\left(\mathbf{A} \leftrightarrow \boldsymbol{n} \right) \overset{\alpha}{\longleftrightarrow} \left(\boldsymbol{n} \leftrightarrow \mathbf{B} \right),$$

$$\left(\mathbf{A} \leftrightarrow \boldsymbol{n} \right) \not\equiv \left(\boldsymbol{n} \leftrightarrow \mathbf{B} \right)$$

In terms of structures as sets, we are saying

$$\mathbf{A} \not\equiv \mathbf{B},$$

$$\left(\overset{=}{\longleftrightarrow} \right) \cap \left(\mathbf{A} \leftrightarrow \boldsymbol{n} \right) = \boldsymbol{n},$$

$$\left(\overset{=}{\longleftrightarrow} \right) \cap \left(\mathbf{B} \leftrightarrow \boldsymbol{n} \right) = \boldsymbol{n}$$

If the two objects $\mathbf{A}$ and $\mathbf{B}$ *are* numbers, then the equation structure simply becomes trivial, as $\mathbf{A} \overset{=}{\longleftrightarrow} \mathbf{B}$, but then, also the two number must obviously be equal. For this reason we might want to say $\mathbf{A} = \mathbf{A}$, for the numbers, but as structures, two equal numbers $\mathbf{A}$ and $\mathbf{B}$ are still $\mathbf{A} \not\equiv \mathbf{B}$. As structures, we have $\mathbf{A} \leftrightarrow \mathbf{B}$, and if they are *two* instances of the same number, then we could say instead, if we wanted to, $\mathbf{A} \leftrightarrow \mathbf{A}$.

When **A** is a function then the nontrivial *equation structure* applies, and the = object is a substructure of both structures it relates as mathematically equal. If $y = f(x)$ then as structures we have the *equation structure*

$$\left(f \leftrightarrow x \xleftrightarrow{\;=\;} x \leftrightarrow y \right),$$
$$f \leftrightarrow \left(x \xleftrightarrow{\;=\;} x \right) \leftrightarrow y$$

in which the = object is a substructure of both the y and $f(x)$ structures, and we also have that third relation

$$(f \leftrightarrow x) \not\equiv (x \leftrightarrow y)$$

which arises because the structure $(f \leftrightarrow x)$ is *not* a number structure, and $(x \leftrightarrow y)$ *is* a number structure.

In terms of a morphism, we have the functional morphism $x \xrightarrow{\;f\;} y$, to which the same *equation structure* applies:

$$x \xrightarrow{\;f\;} y \equiv \left(f \leftrightarrow x \xleftrightarrow{\;=\;} x \leftrightarrow y \right),$$
$$x \xrightarrow{\;f\;} y \equiv f \leftrightarrow \left(x \xleftrightarrow{\;=\;} x \right) \leftrightarrow y$$

We can see here the precise structural way in which a functional morphism f from x to y is a structure composed of the relation between the structure of f and two instances of the x number object, one of which gets changed into another number object by f.

When we have an equation $y = f(x)$ we are dealing with an already existing duality of structures and numbers. But what we are looking at in an equation is an *expressing* of a number duality (one or more instances of the same number object) that arises *through* a morphism. That *expressing* is made through the concept of an "equality" or "equation", but what the concept of the equation *represents*, is *the existence of a number duality created through different structures*.

This principle extends to the more general concept of a morphism between unspecified objects or objects other than numbers. In the mathematical, functional morphism, there is a "common substance" in the domain and codomain which is *Number*. In the more general playing field of morphisms between objects there must also be a "common substance" in the morphism.

Morphisms between different object types

There can be morphisms between different types of objects. In the case of a function $f(x_1, x_2 \cdots x_n)$, f is a structure with inputs and outputs, consisting of rules applied to the inputs, to produce the outputs. In general, the type of object subject to morphism is a number.

Beyond ordinary mathematical functions it is possible for the data of the rules that constitute a morphism, to itself undergo morphisms, *i.e.* changes to the rules, in response to input data. This is a principle employed in artificial intelligence systems. It is also possible for the data of rules to undergo morphisms into the data of outputs.

In these cases the lowest level "common ground" for all structures and forms is binary data, but there are other levels of meaning and structure above, that also form "common ground". In the more general playing field of morphisms the "common ground" is the structure of rules that allows the morphism, occurring at the appropriate level.

Duality and non-duality

The morphism $\mathbf{A} \underset{\rightarrow}{\overset{M}{\longleftrightarrow}} \mathbf{B} \underset{\leftarrow}{\overset{M}{\longleftrightarrow}} \mathbf{A}$ is two instances of $\mathbf{A}$ morphing to one instance of $\mathbf{B}$. We might see this in mathematical structures, for example where it turns out that

$$z = f(x),$$
$$z = f(y),$$
$$x = y$$

If also in this same structure we now have that $\mathbf{B} \underset{\rightarrow}{\overset{N}{\longleftrightarrow}} \mathbf{A}$, then we will have $(\mathbf{A} \leftrightarrow \mathbf{A}) \underset{\rightarrow}{\overset{P}{\longleftrightarrow}} \mathbf{A}$, which is a morphism from two instances of $\mathbf{A}$ to one instance of $\mathbf{A}$.

Many more involved examples exist. In the structure $\mathbf{A} \leftrightarrow \mathbf{B}$ both the objects and the interval may be subject to morphisms. We can have a morphism

$$\mathbf{A} \xleftrightarrow{\; M \;} \mathbf{B} \equiv \mathbf{A} \xleftrightarrow{\; M2 \;} M \xleftrightarrow{\; M3 \;} \mathbf{B}$$

where $M2$ and $M3$ can be morphisms between object types. Suppose M is subject to the morphism $M \xleftrightarrow{\; M4 \;} \mathbf{A}$. Then because we now have

$$\mathbf{A} \xleftrightarrow{\; M2 \;} \left(M \xleftrightarrow{\; M4 \;} \mathbf{A} \right) \xleftrightarrow{\; M3 \;} \mathbf{B}$$

there is a morphism $\omega \equiv M4 \xleftrightarrow{\; \omega \;} M3$:

$$\left(\mathbf{A} \xleftrightarrow{\; M \;} \mathbf{B} \right) \xleftrightarrow{\; \omega \;} \left(\mathbf{A} \leftrightarrow \mathbf{A} \right).$$

We can also say this is

$$\left(\mathbf{A} \xleftrightarrow{\; M \;} \mathbf{B} \right) \leftrightarrow \omega \leftrightarrow \left(\mathbf{A} \leftrightarrow \mathbf{A} \right).$$

In the case that $\mathbf{B} \equiv \mathbf{A}$, so that we are now dealing with two different instances of $\mathbf{A}$, then we would begin with

$$\mathbf{A} \xleftrightarrow{\; M \;} \mathbf{A} \equiv \mathbf{A} \xleftrightarrow{\; M2 \;} M \xleftrightarrow{\; M3 \;} \mathbf{A}.$$

We are now going to number the instances of $\mathbf{A}$ in order to show what is going on, so this becomes:

$$\mathbf{A}_1 \xleftrightarrow{\; M \;} \mathbf{A}_2 \equiv \mathbf{A}_1 \xleftrightarrow{\; M2 \;} M \xleftrightarrow{\; M3 \;} \mathbf{A}_2.$$

We have temporarily changed the convention now, so that different instances *must be* distinguished by different subscript numbers, hence

$$\left(\mathbf{A}_1 \leftrightarrow \mathbf{A}_1 \right) \equiv \left(\mathbf{A}_1 \not\leftrightarrow \mathbf{A}_1 \right) \equiv \mathbf{A}_1$$

which appears clumsy, but will serves to illustrate what is going on.

Only *different numbered* instances are distinct identical objects. Following the same procedure we carried out the first time, now when $\mathbf{B} \equiv \mathbf{A}$, we will have, putting $\mathbf{A}_2$ where previously we had $\mathbf{B}$:

$$\left(\mathbf{A}_1 \xleftrightarrow{\;M\;} \mathbf{A}_2 \right) \xleftrightarrow{\;\omega\;} \left(\mathbf{A}_1 \leftrightarrow \mathbf{A}_1 \right),$$

which is an expression of a morphism from a duality of instantiations of the object to a single instantiation of the object:

$$\left(\mathbf{A}_1 \xleftrightarrow{\;M\;} \mathbf{A}_2 \right) \xleftrightarrow{\;\omega\;} \mathbf{A}_1$$

Structure space

Essentially then, implicit in the idea of distinct instances of identical objects that are capable of morphism, is the idea of the morphism from a duality of such objects to a non-duality. In the case of a reversible morphism, there can be an expansion of distinct identical instances from a single instance.

In structure notation, for a set of n identical objects we would write the morphism from duality to non-duality as

$$\updownarrow\left(\{\mathbf{O}_n\}\right) \xleftrightarrow[\rightarrow]{\omega} \mathbf{O},$$

or the morphism from non-duality to duality as $\mathbf{O} \xleftrightarrow[\rightarrow]{\omega} \updownarrow\left(\{\mathbf{O}_n\}\right)$.

Consider a structure $\mathbf{S}$ that is a set of identical objects separated by identical space instantiations, that can be specified in terms of nodes and edges as:

$$\mathbf{S} \equiv \updownarrow\left(\{\mathbf{N}_p, \mathbf{E}_q\}\right)$$

where $\mathbf{N}$ are the nodes or objects, and $\mathbf{E}$ are the edges or intervals, and because $\mathbf{S}$ is a fully connected graph, $q = \left(p^2 - p\right)/2$. The morphism ω applied to all the edges of such a structure can be specified with a structural function gadget

$$f_1^n\left(\updownarrow\left(\{\mathbf{N}_p, \mathbf{E}_q\}\right)_{\$}\right) \text{ for finite } n,$$
$$\text{or just } f(\mathbf{S}),$$

when $n \to \infty$. The gadget specifies that the intervals in the structure $\mathbf{S}$ must each be subject to the morphism ω for that interval. Because the morphism is reversible we can specify it in one direction or the other as

$\overset{\mathbf{S}}{f}(\mathbf{S})$ for the morphism from duality to non-duality, and

$\overset{\mathbf{M}}{f}(\mathbf{S})$ for the morphism from non-duality to duality,

Now suppose we specify a single instance of **S** as a single (non-dual) "point". At this stage, this is nothing more than an object name. A new structure

$$\updownarrow(\{\mathbf{S}\}) \equiv \mathfrak{S}$$

that is a continuum of identical distinct objects can then arise by applying the infinite gadget to **S**:

$$\left(\overset{\mathbf{M}}{f}(\mathbf{S})\right) \equiv \mathfrak{S}.$$

This structure $\mathfrak{S}$ that is created from the original single abstract point, we can call a *structure space*. It is an infinite space of identical but distinct instantiations of **S**, whose infinitesimal intervals are the structure space itself. The structure space can then be related to a conventional number space $\updownarrow(\{\mathbf{M}\})$, such as $\mathbb{R}$, or $\mathbb{R}^n$, by then specifying

$$\mathbb{S} \equiv \mathfrak{S} \xleftarrow{\;\chi\;} \updownarrow(\{\mathbf{M}\})$$

where $\{\mathbf{M}\}$ is the set of numbers or objects (e.g. vectors or number sets) used to specify (or give an address to) the points of a conventional space $\mathbb{S}$, through the bijective mapping χ from $\updownarrow(\{\mathbf{M}\})$ to $\mathfrak{S}$. We now have a structural space $\mathbb{S}$ of identical points that are uniquely numbered or addressed by a number structure $\updownarrow(\{\mathbf{M}\})$, and we could go on to refer only to the number structure in order to work out how this structure might behave, and how objects and structures within it might be composed.

The number space, although composed of distinct unique objects, is capable of yielding by itself distinct, identical objects, by virtue of its own duplication. Any two instances of $\updownarrow(\{\mathbf{M}\})$ are identical, but can be superposed offset, with the zero or origin points non-coinciding. There are then two distinct but identical objects $\updownarrow(\{\mathbf{M}\})$, as

$$\updownarrow(\{\mathbf{M}\}) \xleftrightarrow{\;S\;} \updownarrow(\{\mathbf{M}\})$$

where the space $\xleftrightarrow{\;S\;}$ is itself an object $S : S \subset \{\mathbf{M}\}$. The objects are metric spaces, separated by a metric space. We can also specify two identical objects $\mathbf{O}$ of any other suitable kind whose separating space is part of a metric space. In contrast, *structure space* requires no metric space in order to yield distinct identical objects.

Now in the full structure of $\mathfrak{S}$ each interval $\mathbf{S} \leftrightarrow \mathbf{S}$ is

$$\mathbf{S} \leftrightarrow \mathbf{S} \equiv \mathbf{S} \xleftrightarrow{\;\alpha\;} R \xleftrightarrow{\;\beta\;} \mathbf{S}$$

where R is the objectified relation between two adjacent points $\mathbf{S}$, and where every instance of $\mathbf{S}$ now has the relation to $\updownarrow(\{\mathbf{M}\})$ given in the definition of $\mathbb{S}$. The relation χ specifies bijection of nodes from the pre-ordered structure of $\mathbf{M}$ to the nodes of $\mathfrak{S}$, which is a structure $\updownarrow(\{\mathbf{S}\})$. Note that this isn't the same as saying

$$\updownarrow(\{\mathbf{S}\}) \equiv \updownarrow(\{\mathbf{M}\}).$$

The only object that has been instantiated twice in $\mathbb{S}$ is the structure of the fully connected graph implicit in every structure. The structural "difference" between the two objects related by the interval χ, is in *the difference in the nature* of the component objects (called "points" in both cases) from which each is composed.

The space $\mathfrak{S}$ is a structure of distinct *identical* objects with no further specification except the name "point". In contrast the space $\updownarrow(\{\mathbf{M}\})$ is a structure of distinct *unique* objects specified as *numbers*, or *coordinates*, or *vectors*, additionally named "points".

Let's take the case where **M** is a vector space $\mathbb{R}^n$. For the *n*-space the relation R is the relation between two vector objects

$$R \equiv \left(x_1, x_2, \cdots x_n\right) \leftrightarrow \left(x_1 + dx_1, x_2 + dx_2, \cdots x_n + dx_n\right).$$

So the relations α and β are not independent because

$$\alpha \equiv \mathbf{S} \xleftrightarrow{\alpha} \left(R \xleftrightarrow{\beta} \mathbf{S}\right),$$
$$\beta \equiv \left(\mathbf{S} \xleftrightarrow{\alpha} R\right) \xleftrightarrow{\beta} \mathbf{S},$$
$$\alpha \leftrightarrow \beta \equiv R$$

The *intervals* between the points of $\mathfrak{S}$ are now structured by the structure of $\mathbb{R}^n$. Its structure has now *specified* the nature of the intervals in $\mathfrak{S}$. However, this does not change the nature of the objects from which $\mathfrak{S}$ is composed. They are still identical objects, whilst the number objects from which $\mathbb{R}^n$ is composed are each unique. Although the two structures are now coinciding or superposed to create $\mathbb{S}$, the means by which each has arisen is quite different. In terms of sets, $\{\mathfrak{S}\} \cap \{\mathbb{R}^n\} = \varnothing$, and $\mathbb{S} \equiv \{\mathfrak{S}\} \cup \{\mathbb{R}^n\}$.

In the full structure of $\mathbb{S}$ the graph now consists of nodes that are uniquely identified by the elements of $\mathbb{R}^n$. So now $\mathfrak{S}$, which previously consisted of distinct objects whose relations were unknown or unspecified, is now *navigable*. The relations are now known. Without such a specification being applied to it, structure space is not navigable.

Ordinarily, we might not have considered the possibility that in navigating the points of a number space such as $\mathbb{R}^n$, through the rules of its structure $\updownarrow\left(\{\mathbb{R}^n\}\right)$, that we may be simultaneously navigating a structure space not formed from $\mathbb{R}^n$ nor governed by those rules, or dependent upon it. The composition of $\mathbb{S}$ makes this possible.

Navigability

If we adhere to a view of the matter of the universe as being composed of sets of objects whose members are identical but distinct, as is the case with fundamental particles and their states, then we can think of natural phenomena as phenomena in structure space, and as the mathematical analysis of nature as the means of making natural phenomena scientifically navigable.

Using mathematical structures and the knowledge of their relations, we can make objective sense of the structures in nature that we experience and comprehend as natural phenomena. But what we should never overlook is that *that* experience and comprehension is itself a structure in nature. We already met this idea at the end of the last chapter and in the *limitation relation*:

$$\updownarrow\left(\left\{\mathbf{O}_{n}\right\}\right) \leftrightarrow \updownarrow\left(\left\{\mathbf{S}_{q}\right\}\right) \equiv \Omega$$

where $\updownarrow\left(\left\{\mathbf{S}_{q}\right\}\right)$ represents the structure of our own intelligence ad sentient experience of being. Now we meet it again in the relation between structure space and number space. Structure space is the basic concept of a space of distinct but identical objects. Number spaces applied to structure space are what enables natural phenomena in structure space to be studied and understood by us, that is, by the intelligence we are being.

This intelligence has arisen in nature, and is currently believed to be explainable as being based on, or having arisen from, structures based on distinct but identical objects, namely, fundamental particles. This is the current position of physics. And yet the principle of distinct, identical objects, in a structure, and in relation to the world we are studying, has not been penetrated into, in the way we are doing here.

The structure of our own intelligence, especially in relation to brain activity, stands to be understood in terms of mathematical structures, that is, structures understood through number spaces. We could say that in order to rise to a new level of intelligence that sees-through this evolutionary intelligence from above and beyond its structure, and is thus able to change it, we need to go beyond the limitation relation.

In order to do this, we'll need to go beyond understanding our existence in ways that fail to fully recognise the tacit role of structure space in our thought and understanding.

A construction rule

The construction rule is the proposition that *a structure must be formed by instantiating objects of at least one space.* What this means is that when we speak of a structure, each kind of object it is composed of can also constitute a space within the meaning of structure theory. Whilst this initially sounds rather mundane and inconsequential, it has, as we shall see, rather profound speculative consequences.

- A metric is a space, and so are objects that appear "in" a metric, that the metric measures. It may be useful to regard the metric as the space in our description, and the objects it contains as the objects. But this is matter of how we are going about understanding and describing the objects and their relations, that we are concerned with. We could also regard the metric *and* the objects it contains and measures, as a space that forms an interval or relation in a further structure.

In the case of Euclidean objects such as lines and triangles in a Euclidean space $\mathbb{S}$, a set $\mathbb{O}$ of Euclidean objects has the set relation to the space $\mathbb{S}$ that is $\mathbb{O} \subset \mathbb{S}$. However, Euclidean spaces are used to measure all kinds of objects and their Euclidean relations, that are not in themselves Euclidean objects, for example, paintings, elephants and molecules. Such objects are often treated *as though* they are Euclidean objects, when in fact, they are not.

For example, we may use a Euclidean space to measure a set $\mathbb{O}$ of ball bearings and the distances and motions between them. We treat the ball bearings as Euclidean objects, as though $\mathbb{O} \subset \mathbb{S}$. But in fact, the true nature of the ball bearings and their relations is such that only only a substructure of the structure that defines them (or in terms of sets, a subset of the property set that defines them and their relations), can be described by $\mathbb{S}$. All $\mathbb{S}$ can do is to describe Euclidean objects and their Euclidean relations. So we need to use other mathematical spaces to describe the other properties of the ball bearings and their behaviour, such as mass and momentum.

To fully describe the situation we need a space that is composed of all the kinds of conceptual objects and relations that are necessary for the situation's full description. Usually, in a scientific description, we describe only chosen *parts* of this situation, separately, using different metrics, often without fully describing the relations between the metrics.

For example, we might describe some massive bodies and their behaviour using a Euclidean metric, and a metric that measures mass and gravity, and a metric that measures time, as we do in Newtonian mechanics. We are using three different metrics, three different *spaces* (in terms of structure theory), which we can call $\mathbb{S}, \mathbb{M}, \mathbb{T}$, respectively, for distance relations, mass and gravity relations, and time. In this description the relations $\mathbb{S} \leftrightarrow \mathbb{T}$ and $(\mathbb{S} \leftrightarrow \mathbb{T}) \leftrightarrow \mathbb{M}$ are unknown, and not described.

The theory of Relativity is an example of where the missing relations or intervals are "filled in". In Relativity the structure $(\mathbb{S} \leftrightarrow \mathbb{T})$ becomes the *spacetime* metric, and the structure $(\mathbb{S} \leftrightarrow \mathbb{T}) \leftrightarrow \mathbb{M}$ is the one described by Einstein's equation.

In the case of a situation where there is a Euclidean space that "contains" a material object that has some properties of matter that we want to refer to, then we are not dealing merely with a Euclidean space and Euclidean objects. The structure must include whatever objects are necessary to describe the features of the material object's materiality, that we want to talk about. This is seen in concepts such as *phase space*, *system space* and *parameter space*. We can say that

$$\text{If } \exists \mathbf{S} \text{ then } \mathbf{S} : \mathbf{S} \xleftarrow{\ M\ }_{\lhd} \mathbf{S}_\mathrm{p}, \mathbf{S}[k] \subset \mathbf{S}_\mathrm{p}[k]$$

where $\mathbf{S}$ is a structure that is not under consideration as a space, $\mathbf{S}_\mathrm{p}$ is a space that includes all the objects of which those in $\mathbf{S}$ are instantiations, M is a morphism, and $\lhd$ indicates that the number of objects in $\mathbf{S}$ is less than or equal to the number of objects in $\mathbf{S}_\mathrm{p}$. $[k]$ is all possible kinds of objects, and any $\mathbf{S}[k]$ is the subset of $[k]$ from which a structure $\mathbf{S}$ is composed.

If we are just looking at the structure of a space, then

$$\text{If } \exists \mathbf{S}_p \text{ then } \mathbf{S}_p : \mathbf{S} \xleftrightarrow{\;M\;} \mathbf{S}_p, \mathbf{S}[k] \subset \mathbf{S}_p[k]$$

where **S** is some potential structure that could be formed from the space, other than the space itself.

If we are looking at a homomorphism or isomorphism H of a space, from $\mathbf{S}_{p1}$ to $\mathbf{S}_{p2}$ then

$$\mathbf{S}_{p1} \xleftrightarrow{\;H\;} \mathbf{S}_{p2} : \mathbf{S}_{p1}[k] = \mathbf{S}_{p2}[k]$$

and

$$\text{If } \exists \mathbf{S} \text{ then } \mathbf{S} : \mathbf{S} \xleftrightarrow{\;M1\;} \mathbf{S}_{p1}, \mathbf{S}[k] \subset \mathbf{S}_{p1}[k],$$
$$\mathbf{S} \xleftrightarrow{\;M2\;} \mathbf{S}_{p2}, \mathbf{S}[k] \subset \mathbf{S}_{p2}[k]$$

Describing this in words, firstly, in any meaningful structure description, the objects from which the structure is composed, are all objects that are a subset of all possible kinds of objects. These selected objects can also be considered to be objects from which spaces in other structures can be composed. Any structure can therefore be considered a morphism of a space or spaces.

All possible kinds of object is itself an object. This object is the source of objects and spaces in structures. We could also consider it as "the space of all possible kinds of object".

On its own, the construction rule may seem a little abstract. Applied to natural phenomena it implies that firstly, and rather obviously, we can only encounter constructs of natural phenomena made from things that already exist. Secondly it implies that what already exists in the structures we know about, is drawn from the "space" of all objects in which all objects are also "spaces"- within the meaning of structure theory - that relate other objects. The construction rule therefore implies that everything, all structures, in nature, are a morphism of the "space" of every possible kind of object. We will go on to see this space as structure space.

The Structure of the Continuum

Numbers as we write and name them are unique objects that in applied use represent quantitates, and in pure mathematical terms represent interval relations that exist in a *natural number structure*.

We define the *natural number structure* **N** as an infinite structure of unique objects containing one object O_0, (the *zero object*), and in which there is a Hamiltonian path called the *number path*, on which each object is specified as being its degree of separation on the path, from O_0. The *natural number structure* represents the structure that gives rise to $\mathbb{N}$.

The natural numbers can therefore be abstractly defined in structural terms as the objects on the number path of a natural number structure. Of course, we ordinarily need the natural numbers to numerically define a degree of separation in the usual way, but structurally, we don't. We can regard the conceived numeral itself as a *representation* of the degree of separation from a zero object along a number path in a natural number structure.

This may be difficult to envisage because we are already so conditioned with the familiarity of the natural numbers as we have learned and know them, and are already conditioned to count with, according to a favoured number base. We have to remember that natural numbers, as objects with no further furnishings are fundamentally unary, like the intervals in the structure, and that counting by number-base is a construct, and tool, of our own intelligence, and another mathematical structure.

So structurally the natural numbers we write or speak are literally the name labels we give to the objects on a number path, and the order in the natural number set is the order of the number path in the natural number structure.

The numbers must be labelled uniquely, because the degree of separation from O_0 of each object on the number path, is unique. Because the number path through the natural number structure is infinite we get around the problem of needing an infinite supply of unique name labels by using a small finite set of unique name labels, whose cardinality is the number base, and then applying an infinite cycling rule to create all numbers larger than the size of the number-base.

Essentially, then, the objects we *conceive* as the natural numbers and give unique names to, can be created through a structural gadget. The first component of the gadget is the natural number structure **N**, which comes with the *zero* object, and the Hamiltonian path (*number path*) through the structure, beginning at the *zero* object, as already described. We then have the structural device

$$_{S}f_{0}^{\pm\infty}(\mathbf{N}) \leftrightarrow {}_{L}f_{N_1}^{N_B}(\{N\}) \rightarrow \{\mathbb{N}\}$$

in which the first structural function $_{S}f$ operates on **N**, beginning with the *zero* object, along the Hamiltonian path in both directions, infinitely. It pipes each object, together with the data for the direction along the path it lies, to the second gadget $_{L}f$.

The second gadget works in cycles as an infinite accumulating counter, using the data set $\{N\}$ which is the base set of number-names, and using the cycling rule to create new numbers greater than $|\{N\}|$ based on the cycling rule.

The $_{L}f$ gadget cycles through the names in the base set $\{N\}: |\{N\}| = B$, creating new numbers through the cycles, and passes its result on each call from $_{S}f$, to the set $\{\mathbb{N}\}$.

The infinite structural function $_{S}f$ thus creates the set $\{\mathbb{N}\}$ as an infinite set of *number symbol sets*, such as 1, 12, 256, and so on, which can be ordered according to the rules embodied in the cyclic gadget $_{L}f$ used to create them.

We could of course easily just set up a cycling counter gadget to create these symbolic number representations infinitely, without considering the structural function $_{S}f$ or what a number actually is. But in fact, such a gadget would have no meaning if it were not for the originating *natural number structure*, whose Hamiltonian *number path* any such counter's functioning actually represents.

Natural numbers as we know them are actually just these names created in this way, such as "five" or "5", representing specific quantities of objects, but the mathematical relations between them, to which actual relations between quantities of objects in the world conform, are the intervals and interval relations in the natural number structure.

Progressing along the number path from the zero object to infinity, we have

$$N_1 \xleftrightarrow{\;R[1,2]\;} N_2 \xleftrightarrow{\;R[2,3]\;} N_3$$

for the structure of three adjacent objects on the path. One of the properties of the intervals changes progressively along the path as the ratio between the the two number-objects adjacent on the interval. As the position on the path $\rightarrow \infty$, so $R[n, n+1] \rightarrow 0$. Also, the first ratio in interval $R[0,1]$ is an escape to infinity.

Another property of each interval is what we know as the *arithmetical interval* or ordinary arithmetical *difference* between two adjacent number-objects, which remains unchanged through the path.

Slicing up an interval

The usual way of envisaging creating a continuum of infinitesimals between adjacent natural number objects, often referred to in discussions concerning infinities and sets, is to recursively divide up the interval between two adjacent number objects. Some Cantor-style arguments work on this basis.

We can create a gadget to do this job for us. Let

$$f_1^{\infty}(N_1, N_2)$$

be a structural infinite iteration function that starts by replacing the interval between the two number objects N_1, N_2 with an object and two new intervals, and goes on recursively to replace all new intervals formed from its action, with objects and new intervals in the same way. This is just structural speak for the ordinary Cantor argument for progressively slicing up the arithmetical interval between two numbers, infinitely. We then find that the gadget is the morphism

$$f_1^\infty(N_1, N_2) \rightarrow \updownarrow_{N_1}^{N_2}(\{\mathbf{O}\}) \equiv \Gamma$$

where Γ is an infinite Hamiltonian path through an infinite structure $\mathbf{O}$ between N_1 and N_1. Along the path the infinitesimal intervals and objects have a structural summation that is the structural interval of $N_1 \leftrightarrow N_2$.

The myth of infinite bijection

The usual Cantor argument states effectively that in the infinite division of the intervals between natural numbers there is no one-to-one correspondence between the natural numbers and the divisions. The infinity of real numbers is therefore said to be an uncountable infinity, and therefore an infinity of a different kind to the infinity of natural numbers.

There is the presumption in the mainstream study of infinite sets that an uncountable infinity is different to a countable infinity, a conclusion arrived at without really understanding the nature and structure of counting, and the nature and structure of numbers. The infinitude of reals is taken in the mainstream to be larger than the infinitude of naturals. These constructs arise from naive objectivity.

Naive objectivity is the presumption made by the mind that what it contemplates as *object*, has no intrinsic connection to or arising from the mind itself that is doing the thinking and contemplating. Therefore it believes it can contemplate the nature of an infinitude and its relation to numbers, without its own unconscious methods and operations of contemplation and thinking being part of the problem. Especially, it is erroneously convinced that by using calculation, based on algebraic manipulations that work in representing the material world as an object, these manipulations independently of itself yield the truth about things that cannot appear in its direct experience of the world, such as infinity.

Because something is an object, or contemplated as an object, doesn't mean that it is other than an artefact of one's own mind. This truth applies to numbers, as much as it does to anything else.

A popular misconception, distinct from the formal "proofs", is that because there is an infinitude of real numbers in the arithmetical interval between every pair of integers, there must be more reals than integers. This kind of illusory psychology is a mental conjuring trick the mind plays on itself and others, arises from mistaking an infinitude for a number, and follows the false reasoning that

$$\left(\infty_1 + \infty_2 \cdots + \infty_n\right) > \infty_1 \text{ as } n \to \infty$$

or more simply that

$$\infty_1 \times \infty_2 > \infty_1$$

as might be the case for numbers. It is of critical importance in the study of infinities to understand that ∞ is not a number. Merely saying that there is are thing called a *transfinite numbers* that is not numbers (so why call it them numbers?) doesn't demonstrate that thinking in terms of numbers has been transcended. Because infinity is not a number, the principle of "proving" that infinitudes are different, through bijection between objects called infinite sets, is flawed.

The principle of bijection between "infinite sets" relies on there being a fixed *relation* between the members of one set and the members of the other. If we put two sets $\{a,b,c...\}$ and $\{\alpha,\beta,\chi...\}$ into one-to-one correspondence then in

$$a \xleftrightarrow{R1} \alpha$$
$$b \xleftrightarrow{R2} \beta$$
$$c \xleftrightarrow{R3} \chi$$

$R1 \equiv R2 \equiv R3$. That's what is meant by one-to-one correspondence. Such a fixed relation is possible where there is also fixed relations in

$$a \leftrightarrow b \leftrightarrow c$$

$$\text{and}$$

$$\alpha \leftrightarrow \beta \leftrightarrow \chi$$

In fact, it is *only* possible under that condition. Let us pretend these alphabets we are using are infinite, and that the latter applies to all the elements of each set. If we extend the column downwards and always the relations are the same, then as the LHS $\rightarrow \infty$, the RHS $\rightarrow \infty$ also, and we say the cardinality of the LHS is the same as that of the RHS.

What happens if some or all of these relations are different? This can happen for example where

$$a \xleftrightarrow{\Delta} b \xleftrightarrow{\Delta} c$$

$$\text{and}$$

$$\alpha \xleftrightarrow{\Phi} \beta \xleftrightarrow{\Phi} \chi ,$$

$$\Delta \neq \Phi$$

Then there is no one-to-one correspondence, and what happens is this:

As we extend the column downwards, then as the LHS $\rightarrow \infty$, the RHS $\rightarrow \infty$ also, and mathematicians say that the LHS and the RHS, because they are not in one-to-one correspondence, therefore have different cardinalities, or are different "sizes". Because here we have replaced the numbers with letters, it is easy to see the fallacy.

In fact, the truth is that any two "infinite sets" or finite sets, can have their members, or some of their members, randomly or serially related one-to-one, with no fixed relation in the correspondences. If the elements *are* ordered numbers but with no fixed cross-set relations, then moving through the correspondences serially, the sequence of numbers in one set can grow at a different rate to the other set, *without this yielding any information on the "size" of the sets.* The "size" of both sets is already specified as unlimited, which is what *infinite* means here, *and it is not a number.*

A little card trickery

Before we go into a more formal explanation, it may be worth illustrating this kind of illusion in terms of a card trick. I tell you I am going to place two decks of cards, deck A and deck B, face down on the table, and I tell you they are both infinite, and different to each other. What you didn't notice is that I simply opened each box labelled "infinite set/deck", and the decks just appeared on the table.

We first turn over the top two cards, to start two new piles, pile C drawn from deck A, and pile D from deck B. We then begin turning over the cards, one deck at a time, into the two piles. The rule is that if we turn over a card onto its pile, and find it is different to the card at the top of the other pile, then we turn over from the same deck again. As soon as we get a matching pair, we swap to the other deck.

We start doing this with deck A, and find that mostly, the cards are different, so we stay with deck A. Every now and then, we get a match, and swap to deck B. Always, when we turn from deck B, we get a matching pair, and have to swap back to deck A. Clearly, the card decks are not randomly shuffled. They are ordered in some way, *and related together*.

At any time, we can count the cards in piles C and D, and always, C will be a larger pile than D. So we conclude that A must be a larger deck than B. This isn't just a guess. We have realised that the decks are ordered in some way, and related together, so we conclude deck A *must* be larger than deck B, even though both decks are infinite.

So now we attempt to grab the whole of deck A and remove it from the table. We manage to remove a chunk of cards but the deck is somehow still there. We try this with deck B and get the same result. Over and over the same thing happens.

It happens because the decks are infinite. We cannot get them off the table unless we jump them back into their boxes, and then they are not on the table any more. So we cannot verify there are more cards in deck A than there are in B, we just convince our self this must be the case. For any two finite decks it *is* the case.

But we are dealing with *infinite* decks. They are magic cards. They only come out of the box, onto the table, where we could start manipulating them, in the first place, by means of a trick. What is this magic? It is the mind itself. The mind that thinks up the cards, thinks they are independent of the mind, and then draws conclusions about them.

The trick began when I said I was going to put two infinite decks on the table. I made it look as though this had nothing to do with the workings of your own mind, when in fact it was *exploiting* the workings of your mind in order create a psychological illusion. Getting the cards onto the table in the first place, is where the trick really is. I didn't really do it. You did. Precisely the same thing happens whenever anyone speaks of an *infinite set*. However, that doesn't mean that the concept of an infinite set is not a useful tool. It is, but as an object, it is merely an object in the mind.

Natural number structure again

If you understand this little parable then you already transcend the trick. But let's go ahead and look at some structures anyway. When we *structurally* analyse a number interval N_1 to N_2 from **M**, which is the natural number path structure of the naturals $\mathbb{N}$, we get a rather different result.

The infinity of objects in the path Γ is created by the structural infinite iteration gadget $f_1^\infty(N_1, N_2)$, whose iterations can be numbered with the natural numbers. For iteration numbers n, the number Q_n of node objects in the structure that forms within the interval $N_1 \leftrightarrow N_2$ at any iteration n, is

$$Q_{n+1} = 2Q_n - 1$$

and the size G_n of the new full structure or graph (intervals plus objects, or edges plus nodes) therefore grows at a rate

$$G_{n+1} = 2\left(Q_n^2 - Q_n\right) + 1.$$

The path Γ has at its two ends N_1 and N_2, which are the *parent nodes* of Γ. The nodes N_1 and N_2 are by definition adjacent nodes in **M**, and consecutive natural numbers in $\mathbb{N}$. The numbers that they are called, are the names given to their degree of separation from **M**'s zero-object. The lower of the two now becomes Γ's zero-object.

The structural function $f_1^\infty(0,1)$ for example gives rise to a structure between numbers 0 and 1, containing an infinite Hamiltonian path Γ beginning on 0. All we have to do is to *specify* the node objects on the path as each being their degree of separation from the chosen zero object, and we have created another natural number structure and Hamiltonian path between the numbers 0 and 1.

For any Hamiltonian number path Γ created by using $f_1^\infty(N_1,N_2)$ and the necessary specification of zero-object for Γ, the node N_1 is redefined as the zero-object of that instance of Γ.

We can just as easily create a gadget

$$_a f_{0,1}^\infty\left(f_1^\infty(N_1,N_2)\right)$$

where $_a f$ applies $f_1^\infty(N_1,N_2)$ with the same specification to each natural number interval (0, 1), (1, 2), (2, 3) *etc.*, in the path **M**, to infinity. We then have a path structure **M** of the naturals $\{\mathbb{N}\}$ that has in each of its intervals between the natural numbers, another set of naturals on a path Γ. However, each path Γ independently redefines its natural number node N_1 in **M**, as its zero-object.

This can be written more simply as a gadget R^∞ that operates on every interval in the naturals:

$$_a f_{0,1}^\infty\left(f_1^\infty(N_1,N_2)\right) \equiv R^\infty\left(\updownarrow\left(\{\mathbb{N}\}\right)\right) \equiv \mathbf{X} \leftrightarrow h\,.$$

The resulting full structure **X** is therefore infinite and fractal, and h is an infinite fractal Hamiltonian number-path through **X**. The situation is now one where in the full structure there is the main natural number path **M**, with a natural number path in-between each pair of its adjacent nodes. This doesn't invalidate the main path, because these new paths are each distinct instances of a natural number path. The main path $\mathbb{N}$ is one object, and each of the other paths Γ are distinct objects, and all identical.

We have used a gadget to create an infinite set of objects (natural number sets) that together, redefine every natural number in **M** as zero, relative to their own structure. The structure and relations in the network **M** remain untouched. The naive temptation is to think that a number is a number and cannot be changed, but actually, a number is whatever you call it, and what characterises numbers is the structure of their relations.

The path h as sets of nodes consists of $\{H\} \leftrightarrow \{\gamma\}$, where $\{H\}$ is the set of natural number nodes along the main number-path, and $\{\gamma\}$ is the set of number paths Γ in the structures in-between those nodes. Remember that we *started* by dividing up the interval between *both* nodes in the main structure (number-objects) that are adjacent on each Γ. For reasons we'll come to, it's best to think of these as each Γ's "parent" nodes.

We began with the gadget

$$ f_1^{\infty}(N_1, N_2) $$

that slices up interval between the adjacent natural number objects N_1, N_2, or parent nodes, by iteratively replacing intervals with objects. This is the process we already met, whereby $N_1 \xleftrightarrow{\ \alpha\ } N_2$ becomes

$$ N_1 \xleftrightarrow{\ \beta\ } \alpha \xleftrightarrow{\ \chi\ } N_2 $$

which becomes

$$ N_1 \xleftrightarrow{\ \delta\ } \beta \xleftrightarrow{\ \varepsilon\ } \alpha \xleftrightarrow{\ \phi\ } \chi \xleftrightarrow{\ \varphi\ } N_2 $$

and so on.

This process (just like the usual Cantor-style way of dividing up an interval) iteratively begins with *two* points, and creates a point in-between. Then the process bifurcates and repeats. The operation/process is actually a *binary tree*, the number of new intervals formed on each iteration number n being 2^n. The infinity that is formed by the "process" isn't formed linearly. There is no "last point" or "end point" because the growth to infinity is rather like a population growth in which reproduction takes place with all members of each new generation being born perfectly simultaneously.

End points and interpretation

Unlike a population growth, the infinitude that the process creates includes all the ancestors, and unlike population growth it is time-independent.

In each path Γ there are two nodes that are also the parent nodes of Γ. For parents N_1 and N_2 we can label the path Γ_{N_1} to indicate the lower parent that is its zero-object. The degree of separation of 1 from 0, in **M**, *but along the path* Γ_0.

On the path Γ_0 we cannot label the node that coincides with the 1 in **M**, "number infinity". It's simply a node in Γ_0 that has an infinite degree of separation on Γ_0 from the zero object of Γ_0, the lower parent node of Γ_0.

The infinity here, which is its infinite degree of separation from the zero object of Γ_0, is not a number. The infinity has been produced *between* the parent nodes, starting with the parent nodes and a first node of Γ_0 placed *between them*. There is no single "last node" in the process of Γ_0's creation, as the number of nodes increases with iteration n of the structural function, as 2^n.

The infinity of nodes produced as $n \to \infty$ must be thought of as laying *between* the two nodes of Γ_0 that coincide with the parents, the 0 and 1 nodes of **M**, just as the infinity of points in a Euclidean line lays between its two end points. The two end points remain the coordinates in the Euclidean plane that they are. There are two end nodes on each instance of Γ, one of which is assigned as the zero node. The other node, whose degree of separation from the zero node is infinite, cannot be assigned a number by that instance of Γ.

Natural number infinities

Let us refer to the Hamiltonian number-path in a natural number structure **O** as $\updownarrow_H \left(\{ \mathbf{O}_n \} \right)$ where the subscript H on the $\updownarrow$ arrow means that the expression refers to a Hamiltonian number-path through the structure.

Putting $\mathbf{O}_n$ rather than $\mathbf{O}$, is done here just to show that the set of objects $\{\mathbf{O}_n\}$ has elements that are unique. As far as *structures* are concerned, each element of $\{\mathbb{N}\}$ is just an object in a structure, an object that is specified as its degree of separation from the zero object. Putting $\{\mathbb{N}_n\}$ is simply a reminder of that specification which makes each member unique. For the whole number-path of an infinite structure, $n \to \infty$.

There is a relation

$$\overset{\text{INSTANCE IN MAIN}}{\left(\updownarrow_H \left(\{\mathbb{N}_n\}\right)\right)} \leftrightarrow \overset{\text{INSTANCE IN } \Gamma_n}{\left(\updownarrow_{\gamma_n} \left(\{\mathbb{N}_n\}\right)\right)}$$

between the naturals as they appear in the main structure and the naturals as they appear in each interval Γ_n. For brevity now let's write this relation above simply as $\mathbf{M} \leftrightarrow \Gamma$. This is the relation between the naturals and an infinity of naturals between each adjacent pair.

The structure Γ pertains to a *continuum* of real number values that exists between the pair of its parent nodes in $\mathbf{M}$. In contrast, the instantiation of the structure $\mathbf{M}$ is not the instantiation of a continuum.

This situation arises because of the different ways in which each structure has been specified. $\mathbf{M}$ is specified simply by instantiation it as the number-path in a natural number structure with no other specifications. Γ is specified as a number path on a structure with a specific method of creation. The method specifies Γ in such a way that Γ is bounded by N_1 and N_2 in $\mathbf{M}$. The important structural difference therefore, between $\mathbf{M}$ and Γ is that $\mathbf{M}$ is *an unbounded infinite natural number path*, whilst Γ is *a bounded infinite natural number path*. $\mathbf{M}$ is an unbounded infinity, Γ is a bounded infinity.

Structurally, a bounded infinite number path between two number objects N_1 and N_2 on the number path of a natural number structure, differs from the unbounded variety (an unbounded instance of $\mathbb{N}$) in that in the *bounded* case only, there is a set of nodes that form an infinite Hamiltonian cycle

$$\mathbf{C} \equiv \overline{\left(N_1 \cup N_2\right)\{\Gamma\}}.$$

In the case of the *unbounded* path we cannot make such a cycle even by instantiating duplicate number objects.

The parent nodes N_1 and N_2 are the *bounding nodes* or *bounding objects* for the bounded natural number path Γ.

The only distinction between **M** and Γ is that Γ has boundary objects, whilst **M** does not. Implicit in the concept of Γ is that it is a substructure of a structure that includes both Γ and its boundary objects. The set of objects on a path Γ and on the path **M** are both the natural numbers $\{\mathbb{N}\}$. However, they are two distinct instantiations, forming a structure $\mathbf{M} \leftrightarrow \Gamma$. For *all* the instances of Γ in **M**, in other words, all the intervals in $\mathbb{N}$ "filled in" each with an instance of Γ, the structure is the relation of **M** to the infinite set of instances of Γ, which is $\mathbf{M} \leftrightarrow \{\Gamma\}$.

Another easy to conceive place where the distinction between a bounded infinity and an unbounded infinity appears, is in the distinction between the infinity of points in an infinitely long Euclidean line, and the infinity of points in the circumference of a circle, or in a line of finite length.

The relation between the bounded natural number infinity $\mathbb{N}_B$ and the unbounded one $\mathbb{N}$ includes the relation between the reals and the naturals. We'll look at this shortly.

Countability and counting

The usual Cantor-style argument is that the reals are "uncountable" whilst the naturals are "countable". This is now widely repeated, but this whole idea of "countable" versus "uncountable" is arguably not even a genuinely mathematical approach, no matter how dressed up in mathematical symbols it might be, but rather, relies on a play on the word *countable*, and the (actually illegitimate) replacement of its normal meaning with the idea of *one-to-one correspondence* (set bijection).

The idea that a set is uncountable is not untenable. However, the idea that this is because its members are numbers that cannot be put in one-to-one correspondence with the naturals, simply confuses the *quantity* of members with a proposed false mathematical and structural relationship between the *values* of the members and the size of the set.

Countable in the ordinary everyday sense, applied to a set of objects, actually means (A) that it is possible to *carry out the act of* counting the objects because the situation allows this, whether or not you complete the task. It also means, (B), that it is in principle possible to arrive at an answer by completing the task.

A "real world" example is: The grains of sand in a large static pile are countable-(A), but probably not countable-(B), if you want to get to lunch on time. The grains of sand on a beach where the tide is breaking are not even countable-(A), because they are moving, although we might be able to estimate. The idea that a set can be *countable* or *uncountable* on grounds that confuse the concept of counting, with the concept of bijection with the set of natural numbers, fails to comprehend the true nature of the natural numbers, and what there relation is, to the act of counting.

In order for any set to be countable-(B), it must first be countable-(A) in some way. However carrying out the act of counting, in mathematics, isn't necessarily what we do in the kindergarten when counting toy bricks. In mathematics it is possible to "count" in many kinds of other ways.

The general concept of *counting* also happens to be of two other kinds, either *time-dependent* (TD), or *time-independent* (TI). (We use the term *time-independent* here, in the same way that it appears in quantum mechanics, rather than suggesting that our intelligence, which engages in these cogitations, is time-independent). TI or TD is relevant to modern physics and computing, and that doesn't mean it's not relevant to pure mathematics.

The problem with simply *defining* the naturals as "countable" merely by virtue of the fact that two instances of the set of them can be put into one-to-one correspondence, which is mathematically trivial, is that this does not in itself mean that the naturals are countable-(B). But it *implies* that they are, *merely* through the *use of the word* "countable" which as a word triggers (inappropriately in this case) its (B)-kind of everyday meaning. Doing this is essentially a piece of mentalism. Defining properties of sets called "countable" and "uncountable" based on such a one-to-one correspondence is therefore not only mathematically naive but also potentially misleading.

Counting is actually a *process*, whilst *quantity* is represented by a number. Counting *results* in a number, but is not in itself a number. Therefore, whether or not a set is *countable*, has no bearing on its size, as a quantity of elements. The Cantorian notion that the naturals are by definition countable, only means they are countable-(A). It is only by means of a piece of mentalism that this is then mistaken for meaning they are somehow countable-(B), or that countable(A) in the case of the naturals is equivalent to countable(B). It is simply not so.

Counting is the process of assigning particular labels or names called "numbers" to any distinct objects related on the number-path of the natural number structure, whose relations, in themselves, have no names, or labels, until we assign them. These "number" labels or names are entirely determined by the language and number-base system for counting, that we choose.

We can represent the natural numbers in their order (of increasing degree of separation from the zero object on the number-path), and their mapping to a set of objects, with a structural gadget:

$$_c f^n_0 \left(\mathbf{M} \xleftrightarrow{\hspace{1cm}} \mathbf{S} \right)$$

where **M** is the unbounded natural numbers on the natural number path, **S** is the structure of a set of objects, and n is the number of steps along the number-path, on which the structural function operates. As $n \to \infty$, $_c f$ will map the natural numbers to the objects of an infinite set. We can achieve the same mapping with

$$_c f^n_0 \left(\Gamma \xleftrightarrow{\hspace{1cm}} \mathbf{S} \right)$$

where Γ is the bounded natural numbers. In both cases these numbers are degrees of separation from a zero object, but not necessarily the same zero object.

For a set of objects to be countable-(A), it is merely necessary for there to be one-to-one bijection, or structural *relation*, between its objects and a natural number path. In any set of distinct objects, there will always be a Hamiltonian path through its structure of relations on which it is possible to assign the conditions for a natural number path. Its objects do not have to have any *particular relation*, fixed or otherwise, to the objects we call *natural numbers*, that we assign to the natural number path.

Any set of objects, and a set of natural numbers we might use to count them with, *are two distinct structures*. This is true even if the first set is a set of numbers. Thus,

$$\{1, 1.2, 1.5, 2, 2.5, 3, 3.5 \cdots n, n+0.5, n+1, n+1.5 \cdots \rightarrow \infty\}$$

is every bit as countable-(A) as if it had not included the spurious 1.2 as the second element. Countability-(A) is also true also of the set

$$\mathbb{S} = \{1, 2, 3, teddy\ bear, 4, 5, x, 6, 7, frying\ pan, 8, 9, \cdots \rightarrow \infty\},$$

even when we specify that $\mathbb{S} : \mathbb{N} \bigcup \mathbb{T}$, where $\mathbb{T}$ is a set of diverse objects that are not necessarily numbers.

The reals

Natural numbers can be generated using an infinite iteration gadget. The gadget requires an input of whatever number-base we choose, and returns an infinite set of numbers by iterating infinitely over the number base. We could, for example, express this as

$$_{\mathbb{N}}f^{\infty}_{B} \rightarrow \mathbb{N}$$

where B is the number-base. The output in the order we usually write such numbers is an infinite set of numbers each in the form

$$N_n B^n \cdots (+) N_3 B^3 (+) N_2 B^2 (+) N_1 B (+) N_0$$

where $N_n < B$ and N_0 is the "units column". The normal "display output" - the number we write - omits the addition signs and parentheses, and omits all instances of $N_n B^n$ that are zero where all instances of $N_n B^n$ further to the left are also zero. It then displays the elements of the series adjunct to each other. We often also adopt the convention of using commas to separate groups of 3 digits from right to left.

The reals between 0 and 1 *in decimal point notation* can also be generated using a gadget. In this case we have

$$_\mathbb{R} f_B^\infty \{_0^1 \to \mathbb{R}$$

and the output, except for 0 and 1, is an infinite set of numbers each of whose values is in the form

$$0.N_1 B^{-1} (+) N_2 B^{-2} (+) \cdots N_n B^{-n} \to \infty.$$

We can deconstruct this number form and find that it is a cyclic function system with infinitely many functions of the finite set of numbers that is the number-base. (We can also do something similar with the gadget for the naturals). The system looks like this:

$$(0 \leftrightarrow .) \xleftarrow{\;\leftarrow \alpha\;} f_0^S (B^{-1}) \xleftarrow{\;\leftarrow \alpha\;} f_0^S (B^{-2}) \xleftarrow{\;\leftarrow \alpha\;} \cdots f_0^S (B^{-n}) \to \infty$$

Each $f_0^S (B^{-n})$ is a conditional cyclic function whose default value is zero, and whose cycle is finite over the numbers of S which is the set of numbers of the chosen number-base B.

On receiving an input trigger from the function on its right, if its value is $\leq B-1$ the function will increase its value by one. If its value is $B-1$ then on this condition it will return to zero value, and output a trigger to the function on its left. When the function adjacent to the $(0 \leftrightarrow .)$ object outputs its trigger, the $(0 \leftrightarrow .)$ object becomes the number one.

This is of course basically the standard mechanical cyclical counter mechanism, also copied in electronic digital displays, except that there is here an infinitude of counter columns (or wheels), for numbers of infinite decimal places.

Essentially the same configuration of cyclic function system can be applied to the interval between any two adjacent natural numbers. The individual cyclic functions are in synchronised relation to each other, the ratios of their cycles arising from the chosen number-base B.

The structure of this device is an infinite *array* of functions (the numeral columns along the horizontal of the written number), and this infinity is an *unbounded* infinity. To reiterate then, the structure of a decimal number (which is to a finite number-base) consists of an unbounded infinity of finite cyclic functions.

Even with infinite counter wheels the counter device can only display discrete numbers, because that is its nature. The irrationals themselves are discrete numbers. For n wheels, as $n \rightarrow \infty$, no mater what the differential between consecutive displayed numbers on the whole counter, whether we consider it to be $d\{B\}$ or zero, or some other concept that appears to be compatible with a continuum, the device can only display a discrete number because that arises from the concept of its construction. We can therefore become falsely convinced that the continuum itself consists of these discrete numbers.

We may like to imagine that the numbers are smoothly changing from one to another at some unseen infinitely right-hand end of the counter or function system, but this is not the objective nature of the system. This discreteness of the values is independent of the unbounded infinity of wheels or cycle functions. Each wheel or cycle function divides the previous (to the left) interval always by the same number-base B, so as $n \rightarrow \infty$ the ratio of the intervals covered by one cycle, between adjacent functions, does not converge, and a continuum condition is not reached.

But in actuality between 0 and 1 we are dealing with a continuum. There is the notion that we can use real numbers expressed through a finite number-base, such as 10, for its description, or that finite number-base expressions of the reals is what the continuum consists of. However, this is mistaken. Indeed, it is true that at any point in the continuum we can find a finite number-base real number, but this does not mean that the real continuum itself consists of finite number-base numbers.

The concept of using of a finite number-base to describe the continuum between 0 and 1, or between any adjacent natural numbers, (or any continuum for that matter), is a false concept. Decimal reals cannot express the continuum even for an infinite number of decimal places.

The infinite number-base

However, if we were using an *infinite number-base* then we would see a different result. The unbounded infinity of wheels in the device is unchanged by the nature of the wheels it uses, or the number of marks around their circumference, which is the number-base B. As $B \to \infty$, we reach the infinite number-base. At infinity there is no need for cyclic number-base functions in order to count naturals as there is now an infinite supply of unique number names. However, if we wish to quantify values between the naturals then we must express fractions of the unit.

If we begin with the finite number-base system then as we increase the number base $B \to \infty$, the unbounded infinity of cycle functions or wheels is unaffected. As $B \to \infty$ the infinity of wheels to the left of the first unit wheel for the natural numbers is redundant and will never turn because the infinity of naturals are now represented on the first wheel. The structure of the infinite number-base numbers therefore consists of an unbounded (at least apparently) infinity (the horizontal *array*) of cyclic functions of *bounded infinities* (the counter wheels), to the right of the first wheel.

Now, rather than a decimal system with a decimal point, we have an infinite series of wheels beginning on the left with the first wheel or function f_1 that rotates smoothly (there are infinite "steps" in its rotation). The entire set of wheels move smoothly as f_1 rotates infinitesimally, through 0 to 1.

The representation of the naturals as a cyclic or bounded infinity here is now inconsistent with our pre-existing notion of the naturals as an unbounded infinity, but for the time being we are not going to worry about this whilst we are focussed on the continuum in one natural number interval only, and we will return to it in due course.

Any "value" on the continuum between 0 and 1 is now described by the state of the whole function system:

$$f_1 \xleftrightarrow{\leftarrow \infty} f_2 \xleftrightarrow{\leftarrow \infty} f_3 \xleftrightarrow{\leftarrow \infty} f_n \cdots \to \infty$$

where now, each f_n is a *bounded infinity* whilst the horizontal *array* is still, as far as we know at this stage, an unbounded infinity. The infinite *array* is now in effect like a "magnifier" for the exact position of f_1 in the continuum of its cycle.

At present, we are envisaging turning wheels, because this is easier, and less abstract. However, later we will see this "turning" simply as a mapping from one bounded infinity to another. Also at present, all the wheels are of course turning in the same rotational direction. Later, we will be discussing "clockwise" and "anticlockwise" mapping as the idea of plus and minus polarity applied to the bounded infinities.

It is possibly easiest now to think of the functions in the array as being "driven" from the left rather than the right. We are currently going to focus on f_1 changing from 0 to 1. Because the number-base is infinite, over a complete cycle of f_2, f_1 moves only infinitesimally, from 0 to 1 in the infinite real continuum. This infinitesimal has further orders of infinitesimal within it, which are also the infinitesimal movements of the other wheels or functions.

As $B \to \infty$ one cycle through the infinity of f_2 corresponds to f_1 ranging over one unit value, which here is 0 to 1. Similarly, over a complete cycle of f_3, f_2 moves infinitesimally over one of its own "unit" values which is an infinitesimal of the interval 0 to 1. So the "units" of f_3

are second-order infinitesimals of the interval 0 to 1, and 3rd order infinitesimals of f_1. And so on.

The device is now more like an analog clock than a digital counter. The "number" that the device as a whole "displays" at any position in the cycle of f_1, is both the exact position of f_1, *and* the positions of all the cyclic functions or wheels as a whole. This is why the device as a whole is essentially just a "magnifier" for what is already the exact position of f_1.

The "number" that the device displays is of course not a number of the kind we are familiar with. It is a *state of the device* that truly represents the continuum between the naturals. The device has been built according to precisely the same rules for building finite numbers of the kind we are already familiar with.

Every state of the device is only possible because of the rotation. Later, we will see that its states arise from mapping between the bounded infinities, rather than actual rotations.

The device as a whole then, is an unbounded infinity of bounded infinities that lays between the naturals, to an infinity of degrees. We can represent this device with the symbol $\odot^{\circ}$. A little thought will show that actually, although we have been thinking in terms of numbers, the device itself, as yet, is *numberless*. Until we actually fix some starting condition for the rotation (mapping between bounded infinities), which will be the "zero-point" for the whole device, then what $\odot^{\circ}$ *is*, and what f_1 *is*, is not actually the real continuum, but the *numberless continuum*.

Every real number is therefore a specific state $\odot^{\circ}{}_x$ of $\odot^{\circ}$, when $\odot^{\circ}$ has a starting point fixed and is the infinite real continuum. There is then a relation $\odot^{\circ}{}_x \xleftrightarrow{N} N_R$ for every real number N_R that is expressed to a finite number-base. Every real number is thus a *translated expression* of a state $\odot^{\circ}{}_x$ in the real continuum, rather than the reals to a finite number-base, themselves constituting the real continuum.

If we consider the whole real continuum including the naturals, then we have that the structure of real numbers as the number-base $B \to \infty$, is $\odot^{\circ}$, to which has been applied a zero-point. If we are going to attempt to handle infinity using algebraic structures that are essentially structures of placeholders for numbers, then we must allow for doing so with the infinite number-base.

We could choose any state $\odot^\circ{}_x$ of $\odot^\circ$ to represent "zero", just as we can choose any node in the infinite graph or structure for the "zero-object" of a natural number Hamiltonian path. So we begin with a zero state $\odot^\circ{}_0$. For the device as a whole, as we have already discussed, we have that the position $P\!f_1$ of f_1 is represented by $\odot^\circ{}_x$, whose state includes $P\!f_1$. This is the relation

$$P\!f_1 \equiv \odot^\circ{}_x$$

for any state of f_1, so as f_1 passes from 0 to 1, $\odot^\circ$ passes from $\odot^\circ{}_0$ to $\odot^\circ{}_1$. For any number-base B the cyclic functions have the cyclic frequency relation

$$^v\!f_n = {}^v\!f_1 B^n$$

so because B and n are always integers, even as $B \to \infty$ and $n \to \infty$ we have $\odot^{[\circ]}{}_x \equiv \odot^{[\circ]}{}_0$, for any x, where here $\odot^{[\circ]}$ is the cycling of the whole device excluding f_1. $\odot^{[\circ]}{}_x$ is the state of this remaining part of the device that represents the continuum between adjacent naturals.

This $\odot^{[\circ]}$ part of the *array* is the unbounded infinity of bounded infinities (the unbounded infinity of wheels), and for brevity we can just call it *the bounded infinities*. They are bounded, nested, infinities of infinities. So we have that *the naturals are the periodicity in the bounded infinities*. If we want to be more precise, we would say *the naturals arising relative to a zero, are the periodicity in the bounded infinities that also arise from the same zero*.

Periodicity of the naturals

The bounded infinities are related to each other in that a cycle of f_2 is a cycle through the continuum between two adjacent natural numbers, whose interval is the first order infinitesimal of the unbounded infinity of the naturals. A cycle of f_3 is then the first order infinitesimal of f_2's cycle, and so on. For $m > n$ a cycle of f_m is the $m - n$ order infinitesimal of the cycle of f_n.

We are looking at infinite bounded infinities nested in the unbounded infinity of the naturals. It may be tempting to think that this means there are different "sizes" of infinity here. However, we cannot force infinities into the moulds of finite numbers. Any attempt to try to conceive the "size" of an infinity is based on the attempt to do precisely that. It must be remembered that $P f_1 \equiv \odot^\odot{}_x$, in other words the state of the whole device is nothing more than a measure of the state of f_1, that is, its position in its passage between two adjacent natural numbers.

The infinity of bounded infinities and the infinity of positions of f_1 in its passage between two adjacent natural numbers, are two different expressions of the same thing. The state $\odot^\odot{}_x$ is just a "magnification" of what is already $P f_1$.

Essentially what we are doing is taking the real continuum and showing that as far as its existence as *numbers* and their construction is concerned, it has the nature of the bounded infinities. It is not any imagined "size" of these infinities that is important here, but the *relation* between these infinities because it is this that determines the periodicity of the naturals in the continuum. The naturals occur with periodicity in the continuum because they occur at the full coincidences of the "zero points" of the bounded infinities, which are periodic coincidences.

The continuum and the naturals

This phenomenon is behind the structure of numbers and number bases, and only when the number-base itself is allowed to go to infinity, is the continuum represented by the structure of numbers. At infinity, the *structure* remains the same, but at infinity the *nature of numbers* associated with the structure is no longer that of discrete objects, but that of the continuum. The structure is that of the bounded infinities, and any of its individual states $\odot^\odot{}_x$, which form the continuum, can be either perfectly expressed by rational numbers or approximately expressed by irrational numbers.

We have that the structure of real numbers $\mathbf{N}$ is a structural function of the number-base B, such that

$$\mathbf{N}(B) \to \odot^{\circ} \text{ as } B \to \infty.$$

But the base B is itself of course a number to a finite number-base, unless $B \to \infty$. So it is possible to describe this situation in another way. We must ask what if the structure of the reals arises from the bounded infinities, rather than the other way around? Then we have the morphism $\odot^{\circ} \to \mathbf{N}$ in which the structure $\mathbf{N}$ of the reals emerges from infinite bounded infinities. We know that this happens simply by instantiating a state $\odot^{\circ}_{0}$ on the bounded infinities, which otherwise, have a state that is the infinite continuum $\odot^{\circ}$ which has no discrete properties. Once $\odot^{\circ}_{0}$ is instantiated, simply by instantiating the concept of zero or nothing as a *state* of the continuum, or point in the continuum, then naturals spontaneously exist in the periodic relative relations between the infinite cyclic functions of $\odot^{\circ}$, and the naturals then appear as discrete states of f_{1} that correspond to the periodic iteration of $\odot^{[\circ]}_{0}$ across all the bounded infinities.

This periodic iteration of $\odot^{[\circ]}_{0}$ that exists in $\odot^{[\circ]}$ can be regarded as an object in its own right, a *structural iteration function* $f_{\mathbb{N}}$ that produces the natural numbers in their native unary form, through infinite iterations, by passing its periodic $\odot^{[\circ]}_{0}$ states to f_{1}, which then becomes $\mathbb{N}$. In other words we can say:

$$f_{\mathbb{N}} \to \mathbb{N}$$

which is the same structure as:

$$f_{1} \leftarrow f_{\mathbb{N}}.$$

The unary naturals are then discrete states of f_{1} within its continuum, and when we create the natural numbers we ordinarily write to a finite number-base, we are taking the states of f_1 and applying a finite cyclic numbering system to them.

What this shows us is that the unbounded infinity of the naturals is an infinity that arises from an infinite *iteration* which we have identified as the structural function $f_{\mathbb{N}}$. It is an infinity of *supply*, the supply of new unary numbers, and an infinity of *periodic states*, of the structural function $f_{\mathbb{N}}$. It is *not* a *quantifiable* infinity, or an infinity of *quantity*. It is better thought of as a process, or function, or morphism, than as a quantity.

Every written natural number *stands* for a quantity, and if it is a unary number, it *is* the quantity it stands for. But the true infinity of the naturals does not lay in the finite-base numbering structure we use to represent them, so there cannot be said to be an "infinite *number*" of natural numbers.

The fundamental reason for this is not that we need to re-name a pre-conceived concept of an "infinite quantity" as "infinite cardinality", or re-name the concept of an infinite number as a "transfinite number". Rather, it is simply that $f_{\mathbb{N}}$, which is the infinity, is in itself not a quantity to begin with, or something that increments, but rather, is an unquantifiable, infinite iteration of a bounded infinity. Only its *output* is a quantity, and this is always finite. Then we try to encapsulate an infinite output of finite number, in the mistaken idea of an "infinite number", we are failing to see through the superficial appearance of finite numbers.

A bounded infinity, such as the points of the circumference of a circle, can become an unbounded infinity simply by instantiating the circle infinitely, but to do so would be arbitrary. In the case of the naturals, we are saying that it is the instantiation of a "zero position" somewhere in the continuum structure, that then spontaneously creates infinite periodic instantiations of states in the continuum that map to the objects we call the natural numbers. This arises because of the nature of the continuum, which fundamentally, is cyclical.

We are saying that the original continuum of the reals $\odot^\circ$ is the prior object, that in itself consists of nested bounded infinities related to each other by infinities, but the "outer shell" f_1 of which is the unbounded continuum of the reals. In this, the periodic naturals arise *if* we fix an *origin state* in the original continuum. We ca refer to this origin state as the "zero point".

An important point to remember about this original continuum is that although in its native state there are infinite bounded infinities, the number of bounded infinities beyond the first two makes no difference to the effective nature of the continuum. One cycle of $\not{f}_3$ maps to an infinitesimal part of the cycle of $\not{f}_2$, whilst one cycle of $\not{f}_2$ maps to the interval between two naturals on $\not{f}_1$, which is an infinitesimal part of the unbounded infinity of naturals appearing on $\not{f}_1$, or the continuum's "outer shell". So infinite cycles of $\not{f}_3$ map to the interval between two naturals. Thus, one cycle of $\not{f}_3$ is the first order infinitesimal of the interval between two naturals.

The outer shell

$\not{f}_1$ in this structure is represented as a bounded infinity, but it is also the case that unless we introduce a further order of infinity, say, $\not{f}_0$, then $\not{f}_1$ is not an infinity that can ever be regarded as completing a cycle, or being cyclically mapped, and is therefore effectively an unbounded infinity, which is the nature of the infinity of the reals as it is usually considered.

Consideration of the infinite number-base and the structural analyses we have been looking at, suggests that the unbounded real continuum infinity may also be considered as the "outer shell" of the structure $\odot^\circ$ that consists *only* of infinite bounded infinities. This happens if this "outer shell" $\not{f}_1$ in $\odot^\circ$ is an infinity that is not *cyclically* mapped to from points in the other bounded infinities. The situation in which the continuum of the reals is mapped cyclically is something we will come to later.

So far we have considered $\odot^\circ$ to consist of an *unbounded* infinity or infinite array of bounded infinities $\odot$. Accordingly we can also regard the infinite array of $\odot^\circ$ as a *bounded* infinity that can "entered" at any point that subsequently becomes the "outer shell" of the structure. This is equivalent to the fixing of a "zero point" $_0\odot^\circ_x$ on a bounded infinity $\odot$.

The bounded infinite array then appears to be, or functions as, an unbounded infinity. Thus $\odot^\circ$ can be considered to be the object created by "entering" or "cutting into" a fundamental, self-bounded infinite object $_\circ\odot^\circ$ that is an infinite *bounded* infinity of bounded infinities.

Thus in $\odot^\circ$ the position of f_1 as the apparently unbounded real continuum infinity, is effectively created by an arbitrary entry point into an infinite *bounded* array. It simply represents the point where we have entered the mathematical object that in itself is already complete and infinite, without end, and without the beginning we create by entering it. Also implied is that the naturals we encounter in the mathematical structures of our world and in our mathematical intelligence, are an artefact of the point of entry into the infinite bounded infinities $_\circ\odot^\circ$.

Thus, the infinity of the naturals is itself, in $_\circ\odot^\circ$, an infinitesimal of an infinity of infinities, and so on. The object $_\circ\odot^\circ$ that embodies these infinities is none other than what arises from the possibilities of mapping between unlimited instantiations of one object - the bounded infinity $\odot$.

Meaning of periodicity in the bounded infinities

One can use some analogy to dispel some possible misconceptions regarding bounded infinities, using circles. The circumference of a circle is a bounded infinity of points. For two circles, O_1 and O_2, where O_1 has the diameter D_1, and O_2 has the diameter D_2, then one order of infinity will appear between the diameters of the two circles, and between their circumferences, when $D_2 \to \infty$. At the same time, the infinitude of points of the circumference of O_2 as $D_2 \to \infty$, remains the same.

Intuitively, we always feel that this cannot be, especially if we are already convinced there are different "sizes" of infinity. But in the case of the bounded infinities it is not so.

The *length* of the circumference of D_2 is πD_2. This length indeed increases as D_2 increases. But this length is not the number of points in the circumference. The relation between Euclidean length or distance or *interval*, and *number of points*, is completely different to the rules governing relations between numbers and Euclidean dimensions. It is arguably as different as the rules of quantum mechanics are from the rules of classical mechanics.

The rules are as follows:

Number of points	Relation direction	Interval
0	>>	no line
1	<<>>	0
finite n	>>	0
infinite	>>	undefined
infinite	<<	$L > 0$

We can see here that behind the smooth continuum of the Euclidean line lies apparently bizarre quantum-like rules. These rules of relation between Euclidean points and lines include the following. The length of a line is always a number. A line of zero length has one point. There is never *any* finite *number* of points in any line >0, only ever *infinite points*, which is not a number. The infinity of points is *not a function* of length, or *vice-versa*. The infinity of points can map to *any* length. Any length maps to the infinity of points.

So as $D_2 \rightarrow \infty$ the infinity of points in the circumference of the circle, being independent of the length of the circumference, is untouched. The Euclidean line, and Euclidean space, are expressions of the real continuum. The Euclidean point is an expression of relations between numbers, or points in Euclidean space. From the "quantum" rules above we can see that there is a pathological-like "function" $E(n,L)$ between the interval L of a Euclidean straight line or distance over a path and the number n of points "in the line", which is always infinite unless $L = 0$.

In a graph of n versus L the function line consists of $n = 1$ at $L = 0$, and a line parallel to the L axis at plus infinity on the n axis, that begins where $L > 0$. The function consists of these two parts.

Consider three circles O_1, O_2 and O_3, where now the diameter of the circles is irrelevant because of the nature of $E(n,L)$. A "zero point" is chosen for each circle, as $_0O_1$, $_0O_2$ and $_0O_3$, so there are then finite intervals along the circumferences:

$$\left(_0O_1\right)\xleftrightarrow{\ \alpha\ }\left(_0O_1\right),\ \left(_0O_2\right)\xleftrightarrow{\ \beta\ }\left(_0O_2\right)\ \text{and}\ \left(_0O_3\right)\xleftrightarrow{\ \gamma\ }\left(_0O_3\right).$$

For a small interval $\delta(\alpha)$ there is then an infinite mapping of points

$$\delta(\alpha)\xleftrightarrow{\ \varepsilon\ }\beta.$$

Similarly, for a small interval $\delta(\beta)$ there is an infinite mapping

$$\delta(\beta)\xleftrightarrow{\ \zeta\ }\gamma.$$

Every point of $\delta(\alpha)$ maps to every point of β, and every point of $\delta(\beta)$ maps to every point of γ. But also, every point of $\delta(\alpha)$ maps to every point of γ. The mapping $_0O_1\rightarrow{_0O_2}\rightarrow{_0O_3}$ which we can call Θ_0, itself is the "zero-point".

As $\delta(\alpha)\rightarrow d\alpha$ there is a mapping $d\alpha\xleftrightarrow{\ \varepsilon\ }\beta$, and we can similarly find a mapping $d\beta\xleftrightarrow{\ \zeta\ }\gamma$. We then have the mapping $d\alpha\rightarrow\beta\rightarrow\gamma$. In $d\alpha\rightarrow\beta$, $d\alpha$ is a "point change" in α from which there are infinite edges mapping to all the points in β. In $\beta\rightarrow\gamma$ there are infinite edges from each point in β, each point in β thus mapping to all the points in γ. We can take every point in each circle, and map from each point to every other point in the same circle, and in the other circles.

In this way the orders of infinitude increase from the circle to the mappings. Even though the sizes of α, β and γ are untouched, *through the choice of mappings* there can be a series of increasing orders of infinity. However, these orders of infinity are apparent, and not something that genuinely indicates different "sizes of infinity".

This choice of mappings from one $\bigcirc$ object to the next, has been made by selectively choosing edges as mappings, from all edges of the fully connected graph of all relations between all points in all the $\bigcirc$ objects. Although we may appear to envisage the mappings geometrically, we are actually leaving the infinity unchanged.

Interpretation

The bounded infinity is as independent of mappings made through it, as the infinity of points in a Euclidean line is, to changes of lengths in the line. We can create different orders of infinity, through infinite mappings, and these different orders of infinity do exist, because we ave made them, but they are merely artefacts of our procedural approach. They exist as *objects*, because they are objects in our intelligence, inseparable from our intelligence itself.

In this way we have seen how a structure of just 3 objects

$$\bigcirc \leftrightarrow \bigcirc \leftrightarrow \bigcirc$$

in which each object is a bounded infinity of objects (points), allows infinite orders of infinity to exist as mappings or relations between the infinity of objects from which the $\bigcirc$ objects are composed. This property is inherent in the infinite structure, or the infinite fully connected graph.

The number of edges R in the fully connected graph (no loop edges) is $\left(n^2 - n\right)/2$ which is always greater than n. So the number of relations between objects in a set in always greater than the size of the set. But as $n \to \infty$ we cannot say $R \to \left(\infty^2 - \infty\right)/2$ no matter how we might express this by other means that look more acceptable. For any infinity, there is no such thing as as an ordinary mathematical function of it that provides a morphism to another infinity. *Any* approach, however it is taken, that effectively tries to say

$$f_1\left(\infty_1\right) = f_2\left(\infty_2\right),$$
$$\left|\infty_1\right| \neq \left|\infty_2\right|$$

is mistaken, by failing to see at least one artefact created by our own intelligence. The theory of infinite cardinalities always follows this form, but never expresses itself in this way, because to do so would be to expose the artefact.

Infinity is not a *number* of objects, but is itself an object with its own properties, that are different to the properties of a number, or an object

91

that is composed of a finite number of objects. As an object it may of course be multiplied, that is, not in the ordinary mathematical sense, but instantiated and re-instantiated, without limit, or infinitely, without changing the object it is. This is true of the instantiation of any identical objects, but in the case of the object of infinity, there is another, additional rule.

Ordinarily, when two identical objects are instantiated, there is a relation $\mathbf{O}_1 \xleftrightarrow{R} \mathbf{O}_2$ in which R is another object such as a space. In the case where $\mathbf{O}_1$ is an infinity, then we have that $\mathbf{O}_1 \equiv \mathbf{O}_2$, and R is an artefact, created in our intelligence, a way of diffracting $\mathbf{O}_1$ into the *appearance* of a duality of $\mathbf{O}_1$ and $\mathbf{O}_2$ through the use of furnishings. In this way the infinite object can be made to appear as multiple or infinite infinities whist remaining in itself unaltered. The "trick" lays in the use of R.

This is why an infinity of infinities is still infinity. If we have an infinite set A, which as an object is a concept, then we can play this game. In order to try to increase a supposed "size" or cardinality of A, we might instantiate another infinite set B and say $A \subset B$. This is a common approach in Cantorian theory, with the intention of "proving" there are different "sizes" or cardinalities of infinity.

But in doing so we have instantiated identical but distinct objects common to both sets. This act takes place in our mind, and the objects are mental. The procedure is one of thought, and processes of thought, pretending to be a representation of something objective in the sense that it is real, independently of the thought. $A \cap B$ is itself infinite, so the structure $A \subset B$ now includes the components

$$A \xleftrightarrow{\subset} B, \ (A \cap B) \xleftrightarrow{\equiv} (A \cap B).$$

We have simply instantiated the infinity $(A \subset B)$ twice, and counted it twice, illegitimately.

Where we have two non-identical objects such as O_{green} and O_{red}, which are each infinite objects, then the structure we are dealing with is actually:

$$O \xleftrightarrow{\textit{furnishings}} O$$

where the *furnishings* are another structure providing the "green" and the "red" properties, that provides the distinction between the infinite objects, but the relation $\bigcirc\leftrightarrow\bigcirc$ in which $\bigcirc\equiv\bigcirc$, remains the same. These furnishings could also be "belongs to set A" and "belongs to set B".

All mental working with sets is working with the concept of a set, which is a mental object. It may, when it is finite, properly map to an object in the material world, and be testable in the material world. But when it is infinite, it is merely a mental object. In fact, what it is, as far as the material world is concerned, is an object that is a brain activity. And as far as the mind is concerned, it is just thought. It is not really "higher thought", but is fundamentally just evolutionary, and based on the animal, sentient experience of a finite world.

A close look at the example above of the three circles $\bigcirc_1$, $\bigcirc_2$ and $\bigcirc_3$, will reveal that having instantiated them as three distinct objects, we proceeded to treat the distinctions as additional furnishings given to the infinite relation $\bigcirc\equiv\bigcirc$, where $\bigcirc$ is just the one fully connected infinite graph, or *infinite structure*.

Morphism to Infinity

The concept of the i-ratio

Real numbers, expressed to a finite number-base, are *representations* of points in the real continuum, which to the infinite number-base is all the possible states $_x\odot^\circ$ of the bounded infinities.

As we have seen, in $\odot^\circ$ there is an infinite array of bounded infinities that are mapped to each other, all with infinite morphisms or mappings between them. We can look again at the nature of the relation between unbounded infinity $\odot$ and $\odot^\circ$.

We can instantiate two bounded infinities and call them $\odot_1$ and $\odot_2$. It is then possible, in the way that we have seen, to have an infinite mapping $\odot_1 \xrightarrow{\infty} \odot_2$, or a finite mapping $\odot_1 \xrightarrow{R} \odot_2$, between the cycles of each. A point in $\odot_1$ mapping to every point in $\odot_2$ is an infinite mapping. If each point in $\odot_2$ then maps onwards to $\odot_3$, another instance of $\odot$, then there is a second-order infinite mapping from $\odot_1$ to $\odot_3$.

Because $\odot_1 \equiv \odot_2 \equiv \odot_3$, which are all one and the same object, this mapping actually place within $\odot$. It is a feature possible because of the infinite fractal nature of the bounded infinity, $\odot$. With finite thinking, we may be tempted to want to find a single atomic "set" of points of a particular "size" in the infinite Hamiltonian cycle, but there is not one.

The relation between the concept of a node or point, as an object, and the concept of the *continuum* of the cycle, as an object, is what produces the concept or object of the *infinite fractal nature* of the object $\odot$. We can even write this relation as:

$$(\text{node}) \leftrightarrow (\text{continuum}) \equiv (\text{infinite fractal nature})$$

A single infinite mapping represents *both* an infinite Hamiltonian cycle *and* a continuum, at the same time, as two different ways of representing the the object $\odot$, just as a circle can represent an infinity of circumference points, and a circumference continuum at the same time. The simultaneous representation of the two is what the *infinite fractal nature* is as an object. The infinite mapping is thus complete, and leaves nothing out of the continuum, and yet, at the same time, we can map again, infinite times, which for convenience of representation we do as mappings between separate instantiations of $\odot$.

We are in effect taking different "zoom levels" of $\odot$ and laying them out as the separate instantiations. The fractal self-similarity between levels, however, unlike in most fractal geometry, has the following nature. As we "zoom in" or "zoom out" to a new level, as it were, we find not only does this new level look exactly like the old, but it *is* the old one.

This ability to combine what is essentially the concept of the infinite set, which is the Hamiltonian cycle, with the concept of the continuum, is what enables us to handle the true nature of infinity, with finite thought. It also lies behind the exposing that the continuum and the infinite set $\mathbb{R}$ are in fact two separate and distinct objects.

Even when we are talking about what is conceived as an unbounded infinity, this fractal nature that occurs when we combine the finite concept of point with the concept of the infinite continuum, is what gives the infinity its bounded nature.

We can express this mapping as an *i-ratio* $x:y$, where the ratio is expressed in terms of cyclic mappings of the interval $_x\odot \leftrightarrow {}_x\odot$ round the whole of each bounded infinity object. We can have a finite real number of cycles to one cycle, or infinite cycles to one cycle. In the first case we talk about the interval, in the second case we are talking about the points or nodes.

This interval is not a *length*, or a "size", as in the notion of a cardinality, but is an abstract representation for the whole of the bounded infinity (or indeed the whole of what is conceived as an unbounded infinity). By instantiating the bounded infinity twice or more, we can map from one instantiation to the other, with various i-ratios.

The i-ratio doesn't change the infinity, or assign it a "size". It just provides, if the ratio is $\neq 1:1$, mappings from one point to all of the infinite positions in the cycle of $\odot$, or to a proportion of the *interval* $_x\odot \leftrightarrow {}_x\odot$ as fixed by the i-ratio. Thus the i-ratio can serve both as a way of handling infinities, and finites.

Thus we might say $\odot_1 \xrightarrow{R} \odot_2$, but what we are doing is expressing a mapping within the one bounded infinity $\odot$, because as we already know, when we instantiate dualities of $\odot$, each instantiation in fact represents one and the same object $\odot$.

We must emphasis again that the interval round $\odot$ or the proportion of it assigned by the i-ratio is not a length, any more than infinity is a number. We use the number 1 to signify the passage through *one* infinite Hamiltonian cycle, through all the points or component objects of $\odot$, *or* to signify the concept of one node. Remember this is a continuum, so the intervals between the points can be replaced *ad infinitum* with new points and intervals. We can still consider the cycle as a node path as long as we understand the nodes and edges are infinitely fractal, and we are not imaging that we are assigning any kind of *number* to the infinity.

The i-ratio $\odot_1 \xrightarrow{1:1} \odot_2$ then refers to this cyclic mapping of one cycle to one cycle. If we first map $\odot \to U$, where U is a numeric unit, and have the i-ratio $\odot_1 \xrightarrow{1:m} \odot_2$ where m is any finite number, then $\odot_2$ will map back again as $\odot \to m$ in the same numeric units. The device $\odot_1 \leftrightarrow \odot_2$ made by instantiating $\odot$ twice, then acts as an ordinary arithmetical ratio.

If we map $\odot \to \infty$, then with the i-ratio $\odot_1 \xrightarrow{1:m} \odot_2$, $\odot_2$ will also map back again as $\odot_2 \to \infty$. But if we map $\odot \to \infty$, *and* have the i-ratio $\odot_1 \xrightarrow{1:\infty} \odot_2$, then each node of the infinity of nodes on the Hamiltonian cycle of $\odot_1$ is now being mapped to *all* the nodes in $\odot_2$, which is 1 cycle of $\odot_2$. Thus, in mapping 1 cycle of $\odot_1$ to the object $\odot_2$, 1 cycle of $\odot_2$ is now mapped from 1 cycle of $\odot_1$, infinitely many times.

Hence $\odot_2$ now maps back again, in i-ratio terminology, as a second-order infinity, as $\odot_2 \to \infty^2$. This does not mean "infinity squared", as though infinity is a number on which we can carry out index operations, although we may colloquially refer to it in that way. The index literally means the mapping has been carried out as described. This is the situation that we meet between *three* adjacent functions in the device of the infinite number-base, rather than two, so this is why we normally begin by assuming the mapping $\odot_1 \to 1$.

Note again that whilst we may have $\odot_1 \to \infty$ and $\odot_2 \to \infty^2$, we still have $\odot_1 \equiv \odot_2$. The second order infinity is an artefact of the mapping, that leaves the unbounded infinity untouched.

So usually, we will not want to begin by mapping $\odot_1 \to \infty$, but by mapping $\odot_1 \to 1$. As in the case of the continuum $\odot^\circ$ we can extend this mapping with another instance of the unbounded infinity $\odot$, as

$$\odot_1 \xrightarrow{1:\infty} \odot_2 \xrightarrow{1:\infty} \odot_3$$

where now the mapping $\odot_1 \to \odot_3$ has the i-ratio $1:\infty^2$. The literal meaning of this is that one object or node of the structure of $\odot$ is mapped to an infinite Hamiltonian cycle through all the objects or nodes of a second instance of $\odot$, and then the same procedure is undergone a 2nd time with a 3rd instance of $\odot$. Thus an i-ratio $1:\infty^n$ involves $n+1$ instances of $\odot$. If we wanted to we could also just instantiate this as $\odot_1 \to \odot_2$ with an i-ratio of $1:\infty^2$.

So when used for finite numbers the i-ratio behaves as an ordinary arithmetical ratio, but in the domain of infinities it takes the forms

$$1:1,$$
$$1:\infty,$$
$$1:\infty^n$$

within the syntax of the i-ratio, where n is a natural number. All of these are infinite i-ratios, but the $1:1$ is also a finite i-ratio. It represents two arithmetical intervals in $\mathbb{R}$. We can laterally invert these i-ratios, so in i-ratios $(1:\infty) \equiv (\infty:1)$, where here we are using the $\equiv$ sign because the i-ratio is a *structural* concept. A little later we will also encounter the i-ratios $1:0$, or $0:1$, and $0:\infty$ and so on, which are also in the domain of infinities. In essence, the n can be thought of an artefact that is a notional fractal "zoom level" on the infinity of $\odot$.

The relation between the infinite and finite i-ratios is that any finite i-ratio (or ordinary arithmetical ratio) has the infinite i-ratio $\infty:\infty$, because the continuum of the interval between zero and any real number always maps to an infinite Hamiltonian cycle of points. So the i-ratio between any two such numbers is $\infty:\infty$.

Unbounded and bounded infinity - the fractal continuum

The infinite bounded infinities $_\circ\odot^\circ$ is an object without beginning or end, in which infinite bounded infinities $\odot$ (infinite instantiations of the bounded infinity $\odot$) are mapped to each other with adjacent i-ratios $1:\infty$ assumed as the default. The infinity of instantiations of $\odot$ is itself bounded. Hence, $_\circ\odot^\circ$ consists of an infinite Hamiltonian path through an infinity of instantiations of $\odot$, the edges or intervals of which could be mappings in any i-ratio.

The conceptual object of an *unbounded* infinity requires a zero-point as a state of the object, or a specific node, for its handling (the zero-point doesn't necessarily have to be the number zero).

If a node object x is extended to infinity, then x is a zero-point. In $x = -\infty \rightarrow +\infty$ there is an implied number zero as the zero point. The reals have the number zero as a zero-point, and the continuum of the reals extends to infinity in both directions away from the zero-point.

Intuitively, it may be tempting to think that in the case of the bounded infinity, in a node-by-node extension away from a zero-point something different is happening to what happens in the case of an unbounded infinity. This thought can arise because of the knowledge that the Hamiltonian cycle through the path is closed, beginning and ending at the zero-point.

However, any finite interval or degree of separation that we assign between a point in either the bounded or unbounded path, and the zero-point, is not the same thing as the degree of separation of that point from the zero-point, *through the structure of the continuum itself.*

As we have already seen, for n nodes the relation between the infinity $n \rightarrow \infty$ and finite numbers in the form $L = f(n)$ is not one of an ordinary mathematical function. If $n \rightarrow \infty$ then there is no smooth morphism $L \rightarrow n$ or $n \rightarrow L$, but rather, the relation follows the pathological-like "function" form $E(n,L)$, where E is "quantum-like" in its behaviour.

Accordingly, and perhaps counterintuitively, because the Hamiltonian path itself is *infinite*, it *does not contain* any finite degrees of separation between *any* of its nodes. Hence every node in the path except the zero-point itself is an infinite degree of separation from the zero-point. This is most easily conceived by understanding that the nature of the path in this respect is *fractal*.

What we are dealing with in the case of any conceived infinity in mathematics, is a continuum, or an infinite extension of discrete objects, or both. The continuum of the reals is both. In a continuum all conceived discrete objects from which the continuum is composed are always an infinite degree of separation from each other. In the case of the extension, the degree of separation of objects from a zero-point increases discretely, infinitely. This is possible because discrete objects that are countable (A) are periodic occurrences in a continuum. This situation we have already described through the object $\odot^{\circ}$, the bounded infinities.

In the case of the supposedly *unbounded* infinity, we may extend through finite node numbers in both directions from the zero-point. But even as we do so, the degree of separation *through the continuum* is already always infinite. All we are doing is exploring the periodicity of the naturals, in the real continuum.

So the apparently unbounded infinity of an infinite extension of discrete objects such as the reals, is actually still accounted for by the structure of $\odot^{\circ}$. In other words, the unbounded infinity owes its infinity to the prior bounded infinity of the continuum from which it arises. It is a secondary effect of the continuum, and *its* nature as a bounded infinity. This secondary effect itself arises from the fact that a zero-point of some kind has been introduced.

The highest level generalised expression of this is the introduction of a zero-point into the infinite bounded infinities $_{\circ}\odot^{\circ}$, by "cutting into" it to create the bounded infinities $\odot^{\circ}$, that then constitute the real continuum and the infinite number-base.

This principle that an unbounded infinity is in fact derived from the bounded infinity of a fractal continuum is notable. In the theory of Relativity there arises the possibility that the universe is finite but unbounded, a phrase Einstein himself used. (Einstein, when he said "the universe" was talking about the nature of the spacetime continuum). For example, the surface of a sphere is finite but unbounded, within its own surface. In other words, it is possibly a manifold of some kind.

In the terms of *our* discussion, this is actually a statement that the space-time continuum is a *bounded* infinity. As far as structure theory is concerned the universe is ultimately a bounded infinity *whether or not* it is "finite but unbounded" in Einstein's sense. This simply because a continuum is a bounded infinity, and any apparently unbounded infinity still arises from the bounded infinity of a continuum.

Within the bounded infinity of the universe the quantity of discrete, distinct objects, could be finite, or could be an unbounded infinity. It just depends on the structures we are using to define and describe these objects.

Ultimately, no matter how we encounter objects through our sentient experience, in science these descriptions and the descriptions of the relations between the objects, are mathematical. The way in which we understand the nature of the universe scientifically, depends on the nature of the mathematical structures we are using to describe it.

These mathematical structures, *as far as we understand them through human intelligence is concerned*, are related to the mathematical structures in our brain activity. And our brain, together with the time the activity takes place in, is a structure. It is a structure of objects called neurons and synapses, and action potentials, and other objects, whose relations are called connections, neural pathways, and neural networks.

Escape to infinity

The arithmetical ratio is a three-object structure which is familiar in the form

$$\frac{y}{x} = z, \ \ \frac{y}{z} = x, \ \ y = xz \, .$$

We can represent this as a structure in the form:

$$z \underset{\leftarrow}{\overset{+}{\longleftrightarrow}} y \underset{\rightarrow}{\overset{+}{\longleftrightarrow}} x,$$

$$x \underset{\leftrightarrow}{\overset{\times}{\longleftrightarrow}} z \underset{\leftrightarrow}{\overset{=}{\longleftrightarrow}} y$$

Here, the values of the objects x, y, z are kept "synchronised" by their relations or mappings. If x, y, z are all allowed to take any values in the infinite real continuum, after we give it a "zero-point", then we can show this structure in terms of relations between instances of the bounded infinity $\odot$. The equation then has the structure **R**:

$$\mathbf{R} \equiv \odot_z \underset{\leftrightarrow}{\overset{\odot x}{\longleftrightarrow}} \odot_y \underset{\leftrightarrow}{\overset{\odot z}{\longleftrightarrow}} \odot_x$$

where in the intervals between the three instantiations of $\odot$, the $\odot x$ and $\odot z$ are the i-ratios mapping between the bounded infinities. As $\odot x \to 0$, $\odot z \to \infty$, and as $\odot z \to 0$, $\odot x \to \infty$, both cases being the division by zero in the equation. Here, the division by zero is a part of the structure of the real continuum $\odot^{\circ}$ as expressed to the infinite number-base, which is the infinity of bounded infinities. In the structure above, at the division by zero (in the equation), there is one order of infinity in the mappings between each pair of adjacent bounded infinities, in the same way as occurs in $\odot^{\circ}$.

In this way, if we take for example the case where $\odot z \to 0$, $\odot x \to \infty$, the i-ratios are $\left[\odot_z : \odot_y \right] = \left[\odot_y : \odot_x \right] = 1 : \infty$. The mappings are such that a single point of $\odot_z$ maps to the whole of $\odot_y$, and a single point of $\odot_y$ maps to the whole of $\odot_x$. There are thus infinite cycles of the $\odot_z$ object to one cycle of the $\odot_y$ object, and infinite cycles of the $\odot_y$ object to one cycle of the $\odot_x$ object.

Thus here, if we were to fix zero points $_0\odot_z$ and $_0\odot_y$, as we did before, then a cycle of $\odot_y$ would constitute a passage through the real continuum between adjacent naturals, and $\odot_x$ will constitute the unbounded infinity of the real continuum, and the structure's "outer shell".

Remember that in $\odot^{\circ}$ only f_1 is strictly necessary as the continuum. All other bounded infinities are just "magnifications" on f_1's state or value. We thus have that in the case of division by zero for $y \neq 0$, $\mathbf{R} \equiv \odot^{\circ}$, which is the infinity of the real continuum.

In the equation $y/z = x$ the x is redundant unless it is an object distinct from the (y/z), and even if it is, the equation means that mathematically it is the same. So by saying $\mathbf{R} \equiv \odot^{\circ}$ we are simply saying that division by zero where $y \neq 0$, is mathematically and structurally one and the same as $\odot^{\circ}$.

In the relation as $z = y/x$ it is easy to erroneously suppose that in the graph of the equation, at $x = 0$, z can be thought of as a line extending parallel to the y axis, "at infinity". The error in this conceptualisation can be seen when it is admitted that there is no value or position on the z axis that is infinity or "at infinity". This is simply because every point on the z axis itself, no matter how far the axis extends, is by definition a finite number. An axis may be infinite, but it only ever provides finite number positions. It infinitely provides them. Infinity is an object that is distinct from the object of the axis and its values. In the case of a graph axis, infinity is a furnishing given to the axis object.

So the parallel line assumption, although colloquially useful sometimes, is in fact incorrect, or a misconception arising from creating a spurious mental association between an axis and infinity, and hence a new but invalid concept of infinity as a point on the axis.

Rather, at $x = 0$, z as an object literally *escapes* altogether from representation by the axis system, which is unable to capture it. The axis system at $x = 0$ merely records the point or points at which it has escaped. So there is a morphism $z \xleftarrow{\;x \leftrightarrow 0\;} \odot^{\circ}$ between finite z values and its escape to infinity that happens where $x = 0$. This morphism can go either way, so also at $x = 0 \pm dx$ there is a morphism from $\odot^{\circ}$ to finite values of z.

Let us for brevity call this morphism $M_{\odot}$. If we are using the infinite number-base, then $M_{\odot}$ in the direction of the escape is structurally a morphism to the disappearance of the zero-point in $\odot^{\circ}$ or $\odot$, from the structure of z. In other words, at the escape to infinity there is no longer any finite interval between a zero-point and a point that z maps to. The finite i-ratio has vanished. Rather, z now maps to the whole of $\odot$ with an i-ratio of $1 : \infty$.

If we are using a finite number-base then it is equivalent to the disappearance of the number-base, and its replacement with the numberless $\odot^{\circ}$ or or $\odot$, which means z as a number is replaced by the numberless $\odot^{\circ}$ or $\odot$. So we can say

An escape to infinity in the case of reals, is the escape from the infinity of the reals to the numberless infinity $\odot^\circ$, or $\odot$, through the disappearance of the zero-point.

This also describes how the z axis, although infinite, cannot measure an escape to infinity. To put it yet another way, it is because an escape to infinity is not the same thing as an infinity of supply, or an infinite iteration or cycle. For example, the naturals that arise periodically in the continuum $\odot^\circ$ to which a zero-point has been applied, are as we have described them, an infinity arising from infinite cycles, and are in infinite supply.

We can easily encapsulate their production with a gadget $f_{\mathbb{N}}(\odot)$ that is a structural infinite iteration function that outputs naturals, on each cycle outputting a new natural, one greater than the previous one. $f_{\mathbb{N}}(\odot)$ represents an infinity, but not an escape to infinity. It's output escapes to infinity over infinite cycles, but only over infinite cycles. In contrast, a division by zero escapes to infinity without requiring an infinity of supply or process in order to do so.

Another morphism takes place when both $x = 0$ *and* $y = 0$. In the graph of $z = y/x$, at $x = 0, y \neq 0$, z escapes to plus infinity for positive y and minus infinity for negative y. But at $x = 0, y = 0$ the escape is both ways, to $\pm\infty$, but also, simultaneously, any z value is then a valid solution in $y = xz$, and the object that fulfils all these conditions is the z axis itself.

We can thus consider the function at this point to be the z axis itself, which is the infinite real number space $\mathbb{R}$. This is reflected in the standard interpretation of $0/0$ as undefined. In other words $0/0$ maps to an infinite object, the infinity of the reals, but not to any individual number.

So to reiterate, where $x = 0, y \neq 0$, there is an escape to $\odot$ (together with a polarity indicating the direction of cycle), but where $x = 0, y = 0$ there is a morphism of z to $\mathbb{R}$. These two conditions we can relate as

$$(z \to \odot) \xleftrightarrow[x=0]{\leftarrow z \neq 0, \; z=0 \rightarrow} (z \to \mathbb{R})$$

where the interval arrow is a "quantum-like" morphism from one state to the other, depending on whether $z = 0$ or $z \neq 0$.

The relation between a positive and a negative $\odot^\circ$ or $\odot$ lies in the direction of the mapping cycle through the infinity. So if we now distinguish between positive and negative we can say there is a structure

$$\left(z = y/x\right) \equiv \left(\overset{s}{f}_-\!\left(-\odot^\circ\right)\right) \leftrightarrow \left(\pm\odot^\circ\right) \leftrightarrow \left(\overset{s}{f}_+\!\left(+\odot^\circ\right)\right)$$

in which morphisms from the central object here to the outer ones take place where $x = 0$, $y = 0$ in the equation or function.

Here, $\overset{s}{f}_+$ and $\overset{s}{f}_-$ are the natural structural gadgets through which the function $z = y/x$ produces its values as finite numbers, out of the real infinite continua $+\odot^\circ$ or $-\odot^\circ$ respectively, by fixing a zero-point on $\odot^\circ$.

What is of most interest is the morphism from the central object to the two outer ones. What is clearly happening is that

$$\left(\overset{s}{f}_+\!\left(+\odot^\circ\right)\right) \text{ and } \left(\overset{s}{f}_-\!\left(-\odot^\circ\right)\right)$$

are structural functions of polarised instantiations of the central object $\pm\odot^\circ$, of which they are morphisms. Since all instantiations of $\odot^\circ$, positive or negative, refer to one and the same object, the $\odot^\circ$, whether mapped "clockwise" or "anticlockwise", then it is easy to see that in this structure, the instances of $\odot^\circ$ appear in much the same way as different instances of the same variable in an equation. In other words the structure of $z = y/x$ is essentially a structural function of $\odot^\circ$.

In this way we can see this particular mathematical structure - the simple ratio - is a structure created from the bounded infinities $\odot^\circ$, that contains morphisms from $\pm\odot^\circ$ to other instantiations of $\odot^\circ$ and structural functions of them. It is not difficult to see how this general principle must apply to any mathematical function or function system that is a function or functions of finite real numbers.

We can also see the incidence of infinite singularities - escapes to infinity - in mathematical real functions, as a natural consequence of the fact that finite real numbers and their relations through functions, are derived in the first instance from the numberless bounded infinities, through the creation of a zero-point.

Transformations and morphisms of the bounded infinity

The structure above consists of transformations and morphisms of the object $\odot^\circ$, which is the real continuum. In itself, the object $\odot^\circ$ has no polarity, but rather, polarity is a furnishing applied to it once we start mapping between instances of it. In the structure

$$\left(z = y/x\right) \equiv \left(f_-\left(-\odot^\circ\right)\right) \leftrightarrow \left(\pm\odot^\circ\right) \leftrightarrow \left(f_+\left(+\odot^\circ\right)\right)$$

the $\left(f_+\left(+\odot^\circ\right)\right)$ and $\left(f_-\left(-\odot^\circ\right)\right)$ are just symmetrical transformations and morphisms of $\pm\odot^\circ$. The object $\pm\odot^\circ$ itself is just $\odot^\circ$ that has been instantiated twice, each instantiation with a different polarity applied to it.

A little thought will show that in a broadly similar way we can take any mathematical function $f\left(x_1, x_2 \cdots x_n\right)$ whose variables are placeholders for numbers that map to the real continuum, in other words, any conventional real mathematical function, and interpret it as a structure

$$f_G\left(\odot^\circ\right)$$

where f_G is a generalised symbol for the natural gadget or device that creates the structure of the function (the causal network of the variables and constants) from instantiations of the bounded infinities $\odot^\circ$. The structure will in general be a structure of relations between transformations and morphisms of these instantiations of $\odot^\circ$.

Because the bounded infinities $\odot^\circ$ are quantitively the same as the one bounded infinity $\odot$, we can also say that any conventional real mathematical function is a structure

$$f_G\left(\odot\right).$$

Something from Nothing

We have seen *structure space* $\mathfrak{S}$ as the abstract, infinite structural space of distinct identical objects (points). $\mathfrak{S}$ can be mapped to from $\mathbb{R}^n$, which itself is a space of distinct unique objects, in order to navigate $\mathfrak{S}$. We have also seen the infinite bounded infinities ${}_\circ\odot^\circ$ as the fundamental object from which we can derive the infinite number-base, the real continuum, the periodic naturals, and the infinity of reals in finite number-bases.

Whilst $\mathbb{R}^n \mapsto \mathfrak{S}$, $\mathbb{R}^n$ does not map to the real continuum $\odot^\circ$, despite the common view that the real continuum consists of real numbers. This is because $\mathbb{R}^n$ consists of infinite, distinct *unique* objects, whilst $\odot^\circ$ consists of infinite, distinct *identical* object instantiations. Thus there is a relation $\odot^\circ \leftrightarrow \mathbb{R}^n$ in which $\odot^\circ$ can produce through some gadget, the reals $\mathbb{R}^n$, but $\mathbb{R}^n$ cannot produce $\odot^\circ$.

The necessary gadget is one that is preceded by entering the infinite bounded infinities ${}_\circ\odot^\circ$, which produces the continuum $\odot^\circ$, and places a zero-point in $\odot^\circ$ as we have previously discussed. This is essentially a conception of the idea of quantity or magnitude, with an infinite number-base. The gadget then proceeds to divide the infinite Hamiltonian cycle of each of the individual bounded infinities in $\odot^\circ$, by a *finite* number-base n.

This division is achieved through the use of a finite i-ratio, $1:n$, where n is the number-base, applied to the pairs of bounded infinities. If working in decimal, which for convenience we will describe, it then inserts the decimal point after the entry position in the array, and maps across the i-ratio divisions to create a decimal real number generating device from the continuum. The infinite set of mappings thus produced becomes the infinite set of real numbers.

In this device, the $\mathcal{f}_1$ cycle of $\odot^\circ$ that appears at the entry point is itself divided by the same i-ratio, and the infinity of periodic naturals that appeared in the infinity of its cycle in $\odot^\circ$, has now been mapped to the infinity of the array "to the left" of $\mathcal{f}_1$. This infinity (1), itself has been mapped in the i-ratios $(1):\infty:n$, where the ∞ here is the infinity of the array (which as already discussed, having arisen from $_\circ\odot^\circ$, is also bounded despite its apparent unboundedness).

We can write this action of the gadget as the morphism $\odot^\circ \xleftarrow{\;\;f_R\;\;}_{\rightarrow} \mathbb{R}$, or write the structural gadget as

$$f_R\left(\odot^\circ\right) \equiv \mathbb{R}$$

Thus, whilst we happily engage in exploring mathematical structures and discovering useful things about them, without the need of a deep knowledge of what numbers really are, or of bounded infinity, or of what the deeper meaning of infinite singularities are, the objects and structures with which we have been working are arising in the first instance through objects that are namely, structure space, the infinite bounded infinities, and the relation between them.

A few preliminary philosophical observations

Structure space is as much a part of our conception, and intelligence, the intelligence we are being, as it is an object apparently separate from it. The *limitation relation* Ω that we met earlier already expresses this:

$$\updownarrow\left(\left\{\mathbf{O}_n\right\}\right) \leftrightarrow \updownarrow\left(\left\{\mathbf{S}_q\right\}\right) \equiv \Omega,$$

where, as we remember, $\left(\left\{\mathbf{O}_n\right\}\right)$ is any structure we understand, and $\left(\left\{\mathbf{S}_n\right\}\right)$ is the structure of our own intelligence. Any structure $\left(\left\{\mathbf{O}_n\right\}\right)$ that we understand in *pure mathematics*, is a structure that is understood in terms of some form

$$\overset{G}{f}(\mathbb{R}) \text{ or } \overset{G}{f}(\mathbb{C}) \text{ or } \overset{G}{f}(\mathbb{R},\mathbb{C})$$

where the gadget $\overset{G}{f}$ is whatever it is in nature that creates the structure we understand, as functional relations between members of $\mathbb{R}$ and/or $\mathbb{C}$. Any structure $(\{\mathbf{O}_n\})$ that we understand in *hard-science*, must be understood in terms of some structure

$$\overset{G}{f}(\mathbb{R},\mathbb{C}) \leftrightarrow (\mathfrak{P})$$

where $\mathfrak{P}$ is some natural phenomenon that $\overset{G}{f}(\mathbb{R},\mathbb{C})$ describes. If we put this in the limitation relation Ω we also have that

$$\left(\overset{G}{f}(\mathbb{R},\mathbb{C}) \leftrightarrow (\mathfrak{P}) \right) \leftrightarrow \updownarrow \left(\{\mathbf{S}_q\} \right).$$

In the case of the natural phenomenon of our own brain, $\mathfrak{P} \equiv \updownarrow \left(\{\mathbf{B}_q\} \right)$, and if we regard the intelligence we are using as explainable in terms of the functioning of the brain, then we have

$$\Omega \equiv \left(\overset{G}{f}(\mathbb{R},\mathbb{C}) \leftrightarrow \updownarrow \left(\{\mathbf{B}_q\} \right) \right) \leftrightarrow \updownarrow \left(\{\mathbf{B}_q\} \right)$$

which is literally just a formal expression in terms of structures, of something that should be already rather obvious in post-naive realism. This is just the fact that our scientific understanding of the nature of our brain, through mathematical form and structure, is related to, or limited by, the structure of the functioning of our own brain.

We also have that because

$$\mathbb{R} \equiv f_{\mathbf{R}}(\odot^{\circ}),$$

the bounded infinities are directly related to our understanding. Even before we have explored the structural nature of imaginary and complex numbers, we can see that the relation of the mathematical structures we comprehend in natural phenomena in general, to those in the functioning of our own brain, as intrinsically related to the bounded infinities.

Zero versus nothing

We can consider *structure space*, the concept of the infinite space of identical but distinct objects called points, as a foundational concept, even if a tacit one, for our further theoretical concepts of infinity, spaces, objects, and structures of non-identical objects, in both the mathematical and material realms. As we have seen, structure space $\mathbb{G}$ is distinct from any number space $\mathbb{S}$, the relation $\mathbb{S} \leftrightarrow \mathbb{G}$ being the mapping between the objects of each kind of space, often through the mathematical concept of a point.

In number space a point is a coordinate giving a unique location in the space. Number space is structurally not uniform and homogeneous because the relations between the numbers are not so, except in respect of the arithmetical difference between adjacent ones. That in itself is insufficient for number space to be considered uniform and homogenous.

In contrast, in structure space a point is an object that is identical in all instances, and the relations between the points is the same everywhere, as a fully connected graph. Also, as the edges between the points can be replaced *ad infinitum* with further points and edges, structure space is not only uniformly infinite, but also uniformly infinitely fractal.

The space of the real numbers is also infinitely fractal, but its structure is not uniform, and, to put it colloquially but most simply, its structure is not independent of fractal "zoom level". Unlike structure space, its structure of relations changes depending on where in it we are looking, and what level we zoom into.

Structure space "looks the same everywhere", whilst number space does not. We might be reminded here of the observable universe. In a particular sense, at a large scale, the universe is said to look the same everywhere, and this is tied to its evolution from the Big Bang. In whichever direction we observe the universe, the further we look, the further we are looking back in time towards its one origin in the Big Bang.

The singularity origin of the Big Bang is what has become, through the expansion of spacetime, the physically extended phenomenon of the universe as we know it, in which positions in spacetime are positions in that metric, which is a mathematical function operating on number space, or number spaces.

The spacetime continuum, as it is called, is not the same thing as the mathematical real continuum. It *uses* the real continuum, but is itself a mathematical functional object of relations between mass and spacetime, expressed in Einstein's equation. Nevertheless, as far as our hard science is concerned, the mainstream view of it is that it is a continuum, and a mathematical structure called a metric.

Where is the distinction between this mathematical structure and the material or physical phenomenon? Where is the distinction between *any* mathematical structure and the material or physical phenomenon that behaves according to it? All natural phenomena like the motions of stars and planets, or quantum particles, behave as mathematical structures, but we encounter and experience the phenomena as something more than abstract mathematical structure.

In nature, the bridge between mathematical structure that we comprehend, and what we call natural phenomena, exists in our self, as the condition of the intelligence we are being. Our experience of nature, our comprehension of it, is an expression of that intelligence, a projection of it. And the imagined scenario of a nature that is separate from it, or from which it is separate, that this intelligence can truly comprehend merely through the experience of sentient being and mind that nature produces through a complexus of neurons called a brain, is an expression of that intelligence's natural naivety.

In science and mathematics, in the mind itself, one of the places that it is possible to go beyond that natural naivety is in the understanding of the infinite singularity. It is also there that in the scientific description of the origin of the universe, or black holes, both of which are singularities, the question of the distinction between mathematical structure and material or physical phenomena rears its head. A mathematical singularity in the mathematical object of spacetime is easy to conceive. A singularity in natural phenomena itself, is not so easy, especially when it is one from which the entire universe is said to spring.

In the morphism $N \xleftarrow{\quad M\odot \quad} \odot$ a finite domain N escapes to an infinite codomain. A domain of finite numbers N that is related or mapped through a function, device, or gadget of some kind, $M \odot$, with an input and an output, whose output *does not output into the finite domain and its codomains.*

In the previous chapter we saw that when we are using a finite number-base, then this escape to infinity is equivalent to the disappearance of the finite number-base, and its replacement with the numberless $\odot^{\circ}$, which means z as a number is replaced by the numberless $\odot^{\circ}$. Alternatively, in the case of $0/0$ there is the morphism $(z \to \mathbb{R}) \equiv M\mathbb{R}$. There is therefore the relations or morphisms $M \leftrightarrow M\odot$, and $M \leftrightarrow M\mathbb{R}$, where M is an ordinary morphism to a finite codomain.

In the structure of $z = y/x$ the morphism $M \leftrightarrow M\odot$ only takes place where $x = 0$, which itself can be considered as the morphism from $(x \neq 0) \leftrightarrow (x = 0)$. The morphism or function M, the ordinary function outside its singularity, can be expressed in terms of the relation between three instances of $\odot^{\circ}B$, which is the continuum to which a finite number-base has been applied. Then the equation or function $z = y/x$ has the structure **S**:

$$\left(\odot^{\circ}B\right)_{z} \xleftrightarrow{\;=\;} \left(\left(\odot^{\circ}B\right)_{y} \xleftrightarrow{\;\div\rightarrow\;} \left(\odot^{\circ}B\right)_{x}\right).$$

As long as $\left(\odot^{\circ}B\right)_{x}$ is not in its relation with $\left(\odot^{\circ}B\right)_{y}$ then nothing extraordinary happens at the zero-point of $\left(\odot^{\circ}B\right)_{x}$. It is the relation that leads to the singularity. The interval $\xleftrightarrow{\;=\;}$ in **S** means that

$$\left(\odot^{\circ}B\right)_{z} \equiv \left(\left(\odot^{\circ}B\right)_{y} \xleftrightarrow{\;\div\rightarrow\;} \left(\odot^{\circ}B\right)_{x}\right).$$

So when $\left(\odot^{\circ}B\right)_{z}$ escapes to infinity, the number-base B on the RHS is lost also. So we then have at $x = 0$:

$$\odot^{\circ}{}_{z} \equiv \left(\odot^{\circ}{}_{y} \xleftrightarrow{\;\div\rightarrow\;} \odot^{\circ}{}_{x}\right)$$

where $\odot^{\circ}{}_{y} \xleftrightarrow{\;\div\rightarrow\;} \odot^{\circ}{}_{x}$ is a $1:1$ infinite i-ratio. We could of course also replace the interval here with a "division" object, and new intervals, and so on, but if we do so we will eventually come back to the same place, with an infinite i-ratio.

The zero-point of $\odot^{\circ}_{x}$ is only the finite *number zero* whilst a finite number-base exists on $\odot^{\circ}_{x}$, and then there is a finite i-ratio mapping from $\odot^{\circ}_{x}$ to $\odot^{\circ}_{y}$. This is the situation when y is finite and $x \neq 0$. But from the "point of view" of the zero-point itself, so to speak, where $x = 0$, the whole of $\odot^{\circ}_{x}$ maps to $\odot^{\circ}_{y}$ with an i-ratio $1:1$, and because there is the escape to infinity, apparently paradoxically *it does so as a numberless continuum in which there is no zero-point*, in other words, no x. We could say that when $x = 0$, the zero point itself, the zero point x has somehow vanished.

What is actually happening is not really a paradox, but simply that the zero point has two different meanings, one when $x \neq 0$ in the equation, and another one when $x = 0$ in the equation.

Relating this back to the equation $z = y/x$ then, it means that in the function, when $x = 0$ the *meaning* of "0" changes. When $x = 0$ the structure of the function itself changes the meaning so it is not *the number zero* as part of a finite number-base, but is literally mathematically *nothing* in the sense that it is the point where the number zero has disappeared together with the number-base.

There is no *number* zero for x that the structure of the function itself can express. When $x \neq 0$ the existence of the zero point *as a number* is merely implied, but in actuality it is just the point in $\odot^{\circ}_{x}$ from which no finite mappings to $\odot^{\circ}_{z}$ exist. Mapping $\odot^{\circ}_{x} \rightarrow \odot^{\circ}_{z}$ jumps from being a finite i-ratio to an infinite one, at the zero-point on $\odot^{\circ}_{x}$.

Putting $x = 0$ does not leave the function intact, with x as the *number zero*, but destroys the structure of the finite function, the object of the network of relations between the variables, with a finite number-base, and leaves in its place the object of the bounded, numberless infinity $\odot$. The true meaning of "singularity" in this case, is a single point at which the finite function itself has disappeared.

That point of disappearance would ordinarily be a point in the number space of the x, if x were not already tied into the structure of the function. In an ordinary or isolated instantiation of the number space it is just a number. But number spaces don't exist as untouchable objects on which functions act without affecting the number space.

Nothing exists as an untouchable object on which something acts. Functions don't exist without numbers, and numbers don't exist without functions. No function exists without instantiating number spaces. And number spaces have inherent within them, functions. They are not two separate things, but are structures, the are connected.

When the number space $\mathbb{R}$ is instantiated as the x object in the structure of the function, it is instantiated with a zero-point. That zero-point is in fact the point at which the number space of x disappears, literally into mathematical nothing, due to the presence the function, whose structure the number space is part of. The number space of y is left unaffected, and the number space of z mapped from x also disappears by escaping to infinity.

So what the function *is*, as it is instantiated, is a *structure* of three instantiations of $\mathbb{R}$, as $\mathbb{R}_z, \mathbb{R}_y, \mathbb{R}_x$, that *changes* the nature of $\mathbb{R}_x$ as it instantiates it, replacing the real number *zero* in $\mathbb{R}_x$ with a zero point that maps to a morphism of the structure itself, of the function. We could describe this morphism in terms of a morphism from finite i-ratios to infinite i-ratios. One way we could write this structure with its morphism, for example, is as

$$\mathbb{R} \xleftrightarrow{\quad x=0,\, y=0 \quad} \left(F\left(\mathbb{R}_z, \mathbb{R}_y, f_F(\mathbb{R}_x) \right) \right) \xleftrightarrow{\quad x=0,\, y\neq 0 \quad} \odot$$

where the structural function $F \equiv (z = y/x)$, operates on $\mathbb{R}_z, \mathbb{R}_y$, and $f_F(\mathbb{R}_x)$, and f_F is the structural function that replaces zero in $\mathbb{R}_x$ with a zero-point that maps $F \to \odot$. This is still only a selective representation of the full structure, and doesn't explore the full range of possible states of F.

The i-product

Two instantiations of the bounded infinity $\odot$ can also be related as an *i-product*. When we are dealing with finite intervals around $\odot$, if one cycle of $\odot_1$ maps to m cycles of $\odot_2$, then n cycles of $\odot_1$ maps to mn cycles of $\odot_2$, where m and n can be any real numbers. The finite i-product is then mn.

If we are dealing with infinities, that is, with the relations of points or nodes in the infinite Hamiltonian cycle, rather than with finite intervals round the cycle as real numbers, then the i-product follows different rules. We must think of the mapping between points, rather than finite intervals around $\odot$.

A single point of $\odot_1$ mapped to a whole cycle of $\odot_2$, is an i-*ratio* of $1:\infty$. The i-*product* is arrived at by taking the i-ratio and "multiplying" the two parts of the i-ratio together. So the i-product here is ∞.

Any *finite* number of cycles of $\odot_1$ to *infinite* cycles of $\odot_2$ is an i-product in which we are mixing *finite intervals around* $\odot$, with infinities. Whenever we do this we must revert to infinite i-products, even though we are talking about an infinity of complete cycles. This is because an infinity of cycles maps to one cycle, with the same mapping as one point to a whole cycle of points (as in the infinite number-base device $\odot^\circ$).

Infinite cycles of $\odot_1$ to infinite cycles of $\odot_2$ is an i-product of ∞^2. The relation is the same as that over three instances of $\odot$, from one point of $\odot_1$ to one cycle of $\odot_2$ to infinite cycles of $\odot_3$, such as we met in the infinite number-base device $\odot^\circ$.

Relation between infinities and finites

We saw earlier that infinite i-ratios follow the pattern

$$1:1,$$
$$1:\infty,$$
$$1:\infty^n$$

and so on, so i-products follow the pattern $1, \infty, \infty^2$, and so on. However, the pattern is broken when we encounter i-ratios involving zero, which we will now look at, or in the infinite i-ratio $\infty:\infty$.

Any finite i-ratio m/n not involving zero, also has an infinite counterpart, the infinite i-ratio $\infty:\infty$. The infinite i-ratios $(p:q)\equiv(q:p)$, where p and q can be 1 or ∞ or ∞^n, because the mapping $(p:q)$ and the mapping $(q:p)$ are really both the same mapping. This is because the $\odot$ object that q occurs on is one and the same object $\odot$ that p occurs on. Remember that multiple instantiations of $\odot$ are not distinct objects.

The infinite i-ratio $1:0$ or $0:1$ occurs when one cycle of points of $\odot_1$ is mapped to or from, a single point of $\odot_2$ (think of the infinite number-base device $\odot^\circ$, however, here, we have just laterally inverted the array). We could write this as $\odot_1 \xleftrightarrow{1:0} \odot_2$. *This is the same as* $\odot_1 \xleftrightarrow{\infty:1} \odot_2$ or $\odot_1 \xleftrightarrow{1:\infty} \odot_2$. If we now swap these around, and say $\odot_1 \xleftrightarrow{0:1} \odot_2$, the infinite i-ratio does not change.

The *finite* i-ratio $1:0$ is also the same as the *finite* i-ratio $0:1$. They are one and the same object. It is one *interval* around $\odot_1$ to *no interval* around $\odot_2$, or the other way around. It transforms to *either* the arithmetical ratio $1/0$ or the arithmetical ratio $0/1$, which are two distinct objects.

The rule is that i-ratios can be expressed either way around, and each way is the same object. A *finite* i-ratio has two possible transformations into an arithmetical ratio, one the inverse of the other. *Infinite* i-ratios do not transform into arithmetical ratios. However, *functions* can transform a finite i-ratio into an infinite one, and *vice-versa*.

The function structure we have been looking at is in the form we met earlier:

$$(z = y/x) \equiv f_G(\odot).$$

So the distinction between the arithmetical ratio $1/0$ and its inversion $0/1$ owes its existence to the prior distinction between the two instantiations of $\odot$ created by f_G. The structural function f_G creates the three spaces x,y,z out of $\odot$, mapped to each other through the structure we recognise simply as $z = y/x$. The structure then falls into two regimes, the first being where the infinite i-ratio between the two instances $\odot_y$ and $\odot_x$ is $\infty : \infty$. There, there is also a finite i-ratio. The second regime is where there is an infinite i-ratio other than $\infty : \infty$. Here, there is the singularity. $f_G(\odot)$ creates the mathematical function and its spaces x, y, z as one structure that has all the features of the function, including its singularity.

So the morphism of $\odot$ that f_G creates as the function and its singularity is in two parts that have the forms:

$$f_{GF}(\odot) \rightarrow \left(\overset{i-ratio}{\infty : \infty} \right) \rightarrow \left(\frac{y}{x} \text{ and } \frac{x}{y} \right)$$

and

$$f_{GI}(\odot) \rightarrow \left(\overset{i-ratios}{1 : \infty, 1 : \infty^2 ... 1 : \infty^\infty} \right)$$

where the morphism from $f_{GI}(\odot)$ here is the structure of the infinite number-base continuum, or the device $\odot^\odot$, and because $\odot^\odot \equiv \odot$, then $f_{GI}(\odot) \equiv \odot$.

We therefore have that this natural, structural function f_G, does nothing to $\odot$, but creates the ordinary mathematical function $f(y, x)$ and its inverse, as y/x and its inverse x/y, from the i-ratio $\infty : \infty$, which f_G creates from its domain $\odot$. However, it creates the function still maintaining the relation $f \leftrightarrow \odot$ which is the function's singularity.

f_G is the natural structural mechanism that creates a duality $\odot \leftrightarrow \odot$, (where in actuality $\odot \equiv \odot$) from the bounded infinity $\odot$, and hence from that duality the i-ratio $\infty : \infty$, from whose transformation arises all the real ratios. The transformation is made through the instantiation of the zero-point in the two instances of $\odot$. In other words, the real ratios are an *artefact* created by f_G, from the bounded infinity $\odot$, by it instantiating a duality of $\odot$'s, fixing a zero-point, and hence allowing i-ratio mapping between the two instances.

The zero-point that f_G fixes isn't intrinsically the *number zero*, but is just a fixed state in the relation $\odot \leftrightarrow \odot$, relative to which will arise, through the concept of mapping between the two instances, the infinite periodicity of the naturals, and the fractal continuum between them, in the way already discussed. The *number zero* is *implied* by the periodicity, and may be subsequently labelled "0", or "number zero", and so on, but what it actually *is*, is the object prior to all mapping, which we informally have been calling the zero-point.

So we are talking about a natural object (a structural function) f_G that creates $\mathbb{R}$ and its ratios from $\odot$, through a very basic principle, of *mapping*, or in other words, a basic principle of *morphism*: $\odot \xrightarrow{f_G} (\mathbb{R}, \{R\})$, where $\{R\}$ is the infinite set of all ratios in $\mathbb{R}$.

This object $(\mathbb{R}, \{R\})$ is a full structure **R** that consists of nodes $\mathbb{R}$ and intervals (edges) that are the ratios and their inversions (the intervals are one ratio in one direction, and the inverse in the other) between the reals. So we have that f_G is the natural morphism from bounded infinity, to this structure **R**, as $\odot \xrightarrow{f_G} \textbf{R}$. $\mathbb{R}$ is the set or number-space of the reals, whilst **R** is the *structure* of the reals, or the structure of the real number-space.

Before being acted on by f_G, bounded infinity $\odot$ is a structure space, because it consists of identical distinct objects, the fractal nodes of the Hamiltonian path through its own full structure. So f_G is a morphism from structure space to the structure **R** of the reals, and to the real number-space $\mathbb{R}$.

The nature of f_G, and the ratio 0/0

We have now presented a number of ways in which $\mathbb{R}$ can be seen to be a mapping from $\odot$ back onto $\odot$, or between instantiations of $\odot$, that arises from $\odot$ itself when it is acted on by f_G. The action of f_G begins by setting a "zero-point" on the unbounded infinity. We have already seen that $\odot$'s nature is infinitely fractal, and that the nature of its infinity is such that it accounts for apparent unbounded infinity. But what is the *underlying* nature of $\odot$ and f_G?

We can regard $\odot$ is as the fundamental object of the aspect of nature that is mathematically describable, or of nature when nature is considered scientifically as an object. As such, it is the principle that lies behind all the mathematical structures through which we understand nature.

Hitherto we might have asked the question "what actually *is* a number"? Now we are saying a real numbers are mappings on the unbounded infinity. These mappings we have expressed here in terms of i-ratios. The i-ratios are either *infinite* or *finite*, and the two kinds map to each other through the ordinary mathematical ratio function. The *infinite* i-ratio maps to the *finite* i-ratio through the ordinary part of the function. The *finite* i-ratio maps to the *infinite* i-ratio only through the singularity.

The finite number objects, and the objects ∞ and ∞^n and ∞^∞ and so on, that appear in i-ratios, can all be called all *i-numbers* (not to be confused with imaginary numbers). The i-numbers are the native objects that arise from mappings on $\odot$. They are then translated by f_G into either real numbers or escapes to infinity in function singularities.

f_G does nor only stand for the creation of the ratio function. It scans for the structure that creates all real functions together with the number spaces on which they operate, as complete structures, from $\odot$. There is a particular kind of f_G for each kind of function and its number spaces. So where does f_G come from, in nature? The fact that it is an object does not mean it has nothing to do with our own mind. On the contrary, we should regard it as part of the nature of the intelligence that we are being.

What about $0/0$? We already saw that this arithmetical ratio, which is usually recognised as arithmetically undefined, structurally maps as $(0/0) \to \mathbb{R}$. This arithmetical ratio transforms into the *finite* i-ratio $0:0$ (zero interval cycles maps to zero interval cycles), as we would expect, and also as we might expect into the *infinite* i-ratio $1:1$ (one point of $\odot_1$, which has a cycle interval of zero, and could indeed be the zero-point itself, maps to one point of $\odot_2$). It also transforms, like all finite real number ratios, into the infinite i-ratio $\infty:\infty$. But it is not an *arithmetical ratio* of $1:1$, any more than is $\infty:\infty$. $0/0$ is another singularity, where the function maps back to the infinite i-ratios only.

False arithmetical ratio expressions

We should always recognise that the expressions $0/0$, ∞/n, $\infty/0$, $0/\infty$, and $,\infty/\infty$ when used or interpreted as "arithmetical ratios", are a mis-use of the arithmetical meaning of *ratio*, and actually in that context meaningless. They are only ever pondered as arithmetical ratios because they are arithmetical ratio lookalikes, and there is a failure to recognise that infinity is not a number.

However, the concept and object of infinity, does indeed occur in structures of relation such as ∞/n, $\infty/0$, and $0/\infty$, but when they do, these are not arithmetical ratios. They are i-ratios, and the function has escaped into the codomain of i-ratios.

We need to recognise that when ordinary computations lead into such expressions, there has been a *transformation* from finite ratios to i-ratios, (or from finite products to i-products). Once we are in the realm of i-ratios or i-products there are different rules that apply, just as much as the rules of quantum mechanics are different from those of classical mechanics. The situation is even in some ways somewhat comparable to the state vector reduction or "collapse of the wave function" in physics.

Fractal and cyclic nature of the continuum

In all the constructs we are using the infinite fractal nature of the continuum is essentially the same thing as the cyclical nature of the bounded infinity, and the mappings between instances of $\odot$ is the same thing as mappings between "levels of zoom" in the fractal. The *levels* are *discrete* because we have made them so, in the same way we talk about points or nodes in the continuum. We simply superpose one object, the point or node, on the other, the infinite continuum, which prices the object or concept of the fractal point.

It is not that there is no "smooth zoom" into the fractal, but rather, that *we set up* discrete "zoom levels" through our concept of discrete structure, that arises through our pre-existing idea of duality and distinct objects. In this way we can develop the knowledge of relations between the finite (discrete) and the infinite (continuum). The structural function

$$f_G(\odot)$$

is an expression of this relation, a gadget representing the process - which we have not yet fully explicated - by which it happens.

The zero-point

The zero point is itself infinitely fractal. We can conceive it either as a point (a node on the infinite Hamiltonian cycle) of the bounded infinity $\odot$, or as a state of the bounded infinities $\odot^\odot$. In $\odot^\odot$ the infinite fractal nature of the continuum is represented as infinities within infinities appearing as the array, in which the i-ratios between adjacent instances of $\odot$ is $1:\infty$. In $\odot$ (which is also $\odot \equiv \odot^\odot$) the infinite fractal nature is represented in the nature of the infinite Hamiltonian cycle that constitutes the way in which $\odot$ is conceived for infinite i-ratios.

The infinite Hamiltonian cycle of $\odot$ is only part of a full structure (fully connected graph). The full structure, as we already discussed, is a structure-space. Specifying an infinite Hamiltonian cycle through the structure space $\mathfrak{G}$ does not change it. However, once we specify a zero-point, all the possible mappings on $\odot$ in i-ratios become possible, and the real numbers $\mathbb{R}$ arise, even if only to the unary number-base, or still in the infinite number-base.

f_G introduces the zero-point, and enables the arising of the numbers. That's an objective way of looking at it. However, quantity recognition, counting, counting systems, and number-bases, are of course all features of *intelligence*. They require an organism, and most of them require a brain. We could even say that the conception or object f_G is essentially representative of the principle of the brain, not merely its physical principles, but the principle through which it creates mind, and the experience of being, self, and the natural world.

Naive and post-naive scientific world view

We haven't yet dealt with imaginary numbers, but if we do we'll see imaginary numbers too, in terms of $f_G(\odot)$. If we can bear with this for a moment, then we can conclude now what there is to say about f_G at this point.

We are describing f_G as the natural structures that act on $\odot$ to produce structures we recognise as functions that map to the natural phenomena we encounter, and also map back again to $\odot$. We can even say

$$f_G(\odot) \rightarrow \quad \begin{array}{c} f_{\mathbb{R}}(\mathbb{R},\mathbb{C}) \\ \updownarrow \\ \odot \end{array}$$

where $f_{\mathbb{R}}(\mathbb{R},\mathbb{C})$ is the generalised structural form of any function. The reason $f_G(\odot) \rightarrow$ is a one-way morphism is simply that it is our own scientific intelligence in action, as produced by nature, in seeing the structure of the world in terms of the RHS object $f_{\mathbb{R}}(\mathbb{R},\mathbb{C}) \leftrightarrow \odot$, or perhaps at least seeing it in terms of the object $f_{\mathbb{R}}(\mathbb{R},\mathbb{C}) \leftrightarrow \infty$, if we don't recognise that all instances of ∞ that we encounter in our mathematical descriptions are instances of $\odot$ and its mappings.

In a truly post-naive scientific understanding, we will have to go beyond this stance of believing in the one-way nature of this morphism. We will have to come to recognise that the world we imagine has a mathematical structure $f_{\mathbb{R}}(\mathbb{R},\mathbb{C}) \leftrightarrow \infty$ arising completely independently of a natural structure $f_G(\odot)$ through which we come to be thinking such thoughts in the first place, this world, doesn't exist as an object.

The Mandelbrot Set

Imaginary numbers

Imaginary numbers are an indispensable part of our scientific understanding of nature, and are fundamental to the mathematical descriptions in our most successful theory - quantum theory. And yet they are imaginary. That means that nowhere in nature can we isolate and display as a material quantity, an imaginary number. That in itself ought to be sufficient for us to suspect that any supposed separation of our own intelligence from the objects of nature that we investigate in science, is false.

The structure of a complex number is simply $m \xleftrightarrow{+} \left(i \xleftrightarrow{\times} n \right)$, where m and n are real numbers. The i itself in the imaginary number object, we can structurally describe as $\left((i) \xleftrightarrow{\times} (i) \right) \equiv (-1)$. Essentially, the imaginary number object i is only functional in multiplication with a real number (even if ± 1), to form an imaginary number.

The structural function $f_C(iR_1, iR_2)$ for the multiplying of two imaginary numbers is the morphism

$$\left(\left(R_1 \xleftrightarrow{\times} i \right) \xleftrightarrow{\times} \left(R_2 \xleftrightarrow{\times} i \right) \right) \xrightarrow{f_C} \left((-1) \xleftrightarrow{\times} \left(R_1 \xleftrightarrow{\times} R_2 \right) \right)$$

The structure of computations for complex numbers is the same as for two distinct variables, except that where imaginaries are multiplied the product is formed in this way. So in terms of i-products, in imaginary number products the direction of mapping to the last instance of $\odot$ is reversed relative its direction for reals.

An i-product for two numbers requires three instances of $\odot$, two for the i-ratio and the third for the result. So the i-product for two imaginary numbers can be regarded as a mapping:

$$\left(\odot_1 \xleftrightarrow{i-ratio} \odot_2 \right) \xrightarrow[\text{reverse direction mapping}]{i-product} \odot_3 .$$

The single mapping $z \rightarrow z^2 + c$ has the structure

$$(z) \overset{\rightarrow}{\longrightarrow} \left((a) \overset{+}{\longleftrightarrow} \overset{i}{(b)} \right)$$

where the object $\overset{i}{(b)}$ is an imaginary number. It has the same i-number structure as a real number, except that it is flagged "imaginary". When combined with another imaginary number in a product morphism (multiplication), the presence of the flag causes the reverse-direction mapping to the resulting codomain number (the product), where the flag is dropped.

The infinite iteration function $z \rightarrow z^2 + c$ where c is another complex number, then has the corresponding structure

$$(z) \overset{\rightarrow}{\longrightarrow} \left(\left\{ (x) \overset{+}{\longleftrightarrow} \overset{i}{(y)} \right\} \right)$$

where $\left\{ (x) \overset{+}{\longleftrightarrow} \overset{i}{(y)} \right\} \equiv P$ is an infinite set of complex numbers.

The values of the members of P as complex numbers depends on the starting value z_0 of z, and the value of c. For $z_0 = 0$, P may or may not diverge as the iteration number $n \rightarrow \infty$, depending on the value of c. The values of c for which it does not diverge are the Mandelbrot set. The Mandelbrot set is a useful example of the relation between number space and structure space.

Duality from structure space

Structure space $\mathfrak{S}$ is the infinite space of duality of identical objects. Number space $\mathbb{N}$ is the infinite space of unique objects. $\mathbb{N}$ is already inherent in $\mathfrak{S}$, *relative to the points* of $\mathfrak{S}$. Relative to each point $\mathfrak{S}_0$ the structure of $\mathfrak{S}$ is identical. However, relative to the duality $\mathfrak{S}_0 \leftrightarrow \mathfrak{S}_1$ all the points of the entire structure are unique, because any third point $\mathfrak{S}_2$ of $\mathfrak{S}$ now necessarily has a unique relation to $\mathfrak{S}_0$ and to $\mathfrak{S}_1$.

This unique relation begins as a unique finite i-ratio interval on the Hamiltonian cycle through $\mathfrak{S}$, when $\mathfrak{S}$ is considered as $\odot$. The degree of separation on the cycle, of $\mathfrak{S}_0 \leftrightarrow \mathfrak{S}_1$ or $\mathfrak{S}_0 \leftrightarrow \mathfrak{S}_2$, however, remains infinite, because as we have already seen, the degree of separation of any two points in $\mathfrak{S}$ is always infinite, due to $\mathfrak{S}$'s infinite fractal nature. The finite i-ratio of each is a real number in the interval from 0 to 1, one cycle of $\odot$, when $\mathfrak{S}_0$ is assigned as the zero-point.

$\{n\} \leftrightarrow \infty$ The assigning of the zero-point to a point in structure space $\mathfrak{S}$, is thus all that is necessary in order to turn structure space $\mathfrak{S}$ into the space $\mathbb{R}$ of real numbers, to the infinite number-base. Continued infinite mappings around the cycle $\odot$ as new instances of $\odot$ then create the apparently unbounded periodic infinity of the naturals.

Therefore, $\mathbb{R}$ itself, to the infinite number-base, *is* structure space $\mathfrak{S}$ to which a zero-point has been assigned. It is merely necessary to recognise that $\odot$ is an infinite Hamiltonian cycle through $\mathfrak{S}$.

Conversely, if the zero-point is now *unassigned*, then $\mathbb{R}$ to the infinite number-base vanishes, leaving $\mathfrak{S}$. This is the escape to infinity that occurs from the zero-point when the structure of a function maps the zero-point back to $\odot$, which is the just infinite Hamiltonian cycle through $\mathfrak{S}$.

The infinity of the reals doesn't occur in the numbers themselves. It occurs in the infinity of structure space $\mathfrak{S}$. Any infinite series is essentially an infinite iteration function f_i. We can number the iterations with natural numbers n. As $n \to \infty$, f_i may or may not converge. If it converges, then it remains finite. If it diverges, or does not converge, then it escapes to infinity. The escape only occurs when $n \to \infty$. That is, it only actually occurs when $n \to \infty$ is a morphism from finite number objects n to the infinity object ∞.

The infinity object ∞ is a distinct object from the object that is the set of all number objects $\{n\}$. There is a relation $\{n\} \leftrightarrow \infty$ which is first necessary before there can be a morphism $n \to \infty$.

We have first that $f_i \leftrightarrow n$. Then, if the morphism $n \to \infty$ occurs, then the morphism $f_i \to \infty$ also occurs. So there is the structure

$$\big((f_i \leftrightarrow n) \leftrightarrow (f_i \leftrightarrow \infty)\big) \leftrightarrow (n \leftrightarrow \infty)$$

Now because f_i is a function $f_i(n)$, we can see easily here that if there is the morphism $n \to \infty$ and $f_i \to \infty$, then f_i and n disappears, leaving just the object ∞ and its self-relations. In other words, there *is* an escape to infinity, but only in the *morphism* $n \to \infty$. If we work on the mistaken principle that $n \to \infty$ means "n tends to infinity but remains n", then there is no escape to infinity.

So a diverging infinite iteration only escapes to infinity in the actual morphism of its iteration numbers n to infinity. That morphism is the disappearance of the zero-point in what was $\mathbb{N}$ before its disappearance, which was also the zero-point in $\mathbb{R}$ or $\mathbb{C}$.

So inherent in the principle of an infinite iteration function is that the structure of the function will map back to infinite structure space $\mathfrak{S}$ when its iteration number n undergoes a morphism to infinity, *if* its structure is that of a *diverging* function. If its structure is that of a *converging* function then it remains intact as the finite structure of relations that it creates through instances of $\mathbb{R}$ or $\mathbb{C}$.

The Mandelbrot set

However, divergence and convergence are not the only possibilities. The well known infinite iteration function $f_i(z,c)$ that is $z \to z^2 + c$, where z and c are complex numbers, and which leads to the Mandelbrot set, diverges, converges, remains infinitely finite in value, or creates finite *attractors*, depending on the conditions for z and c.

There are conditions for which $f_i(z,c)$ will escape to infinity if the morphism $n \to \infty$ occurs, but other conditions for which the other possibilities apply. If the initial value z_0 of z is zero, then the behaviour of $f_i(z,c)$ depends on c. There are then infinite unique behaviours, each as a codomain C_c of $f_i(z,c)$.

Each C_c either escapes to infinity when $n \to \infty$, or remains finite when $n \to \infty$, depending on c. Any C_c that escapes a circle of radius 2 around the origin of the complex plane will escape to infinity when $n \to \infty$.

The parameter c is an instance of the complex number-space. So $f_i(z,c)$ applied over all c is a structural function $f_M(c)$, and there is the structural morphism $c \xrightarrow{f_M} C$ where the structural codomain C is the set of all C_c.

So the object C (the infinite mapping of the complex number-space c though the structural function f_M) bifurcates into these two distinct objects, B which consists of values of c that remain always finite even when $f_M \to \infty$, and E which consists of points that escape to infinity when $f_M \to \infty$. B is the Mandelbrot set.

The Mandelbrot set is infinitely fractal, which manifests clearly as infinite fractal geometry around its perimeter. The fact that its infinite fractal nature is easily seen in its perimeter does not mean that it is not infinitely fractal in its interior. The difference between the two is that around the perimeter both B and E occur, whilst in the interior, only B occurs. Nonetheless, structurally, every point of B is fractal.

It is the fractal nature of the points themselves that allows the fractal nature of the object as a whole. The Mandelbrot set aptly illustrates how points in the complex number-space are themselves iterative and fractal. The reason this is brought to light in the Mandelbrot set (and other fractals generated from infinite iteration functions) is that it is generated from an infinite iteration function, which exposes the the infinite iterative nature of the number-space itself created from $\odot^\circ$, and its complex counterpart.

The fact that E is also fractal, manifests in the disconnected Julia sets found outside the Mandelbrot set. The disconnected Julia sets - which outside the Mandelbrot set are all disconnected - are well known to consist of "disconnected points", or "Cantor points". In this case, two adjacent points have the structure $P_1 \leftrightarrow P_2$, where P_1 and P_2 are distinct objects, i.e. $P_1 \not\equiv P_2$.

According to structure theory, that means that there must be a space object separating them as $P_1 \xleftrightarrow{\ \mathbf{S}\ } P_2$. Also, by structure theory, we can expand this into $P_1 \leftrightarrow \mathbf{S} \leftrightarrow P_2$, and then expand again, *ad infinitum*, in the way we are already familiar with.

Each time we do this either **S** is another Cantor point or it is not. Only when there are instances of **S** that are not Cantor points, that is, not number objects in $\mathbb{C}$, is the space of Cantor points disconnected.

Connectedness of $\mathbb{C}$

Any two distinct numbers in the complex number-space $\mathbb{C}$ also have the structure $P_1 \leftrightarrow \mathbf{S} \leftrightarrow P_2$, where the space **S** can be any number in $\mathbb{C}$ that is distinct from P_1 and P_2. We can also specify the condition that $P_1 < \mathbf{S} < P_2$. We can then apply an infinite structural iteration function to this condition that takes the interval $P_1 \leftrightarrow \mathbf{S}$ and replaces the interval to make the structure $P_1 \leftrightarrow P_n \leftrightarrow \mathbf{S}$ where n is the iteration number and P_n is half way between P_1 and **S**. It then swaps the names of **S** and P_n, to make the structure $P_1 \leftrightarrow \mathbf{S} \leftrightarrow P_n$, and then repeats the whole procedure *ad infinitum*.

The arithmetical interval $P_1 \rightarrow P_n$ is always one half of $P_n \leftrightarrow \mathbf{S}$. When $n \rightarrow \infty$, then in the structures there is $(P_1 \leftrightarrow \mathbf{S}) \rightarrow (0)$, and arithmetically $(P_1 \rightarrow \mathbf{S}) \rightarrow 0$ (because $(1/2)^n \rightarrow 0$ as $n \rightarrow \infty$). So in the infinite iterative structural process of replacing the interval in $P_1 \leftrightarrow \mathbf{S}$ with a space object, the space object is always another complex number distinct from P_1 and **S**, and only when $n \rightarrow \infty$ does the space object become the number zero.

We then have the structure $P_1 \leftrightarrow (0) \leftrightarrow \mathbf{S}$ when $n \rightarrow \infty$. The arithmetical condition we set, that the central object here is arithmetically half way between two outer distinct number-objects, has been transcended by the escape to infinity of the iteration function. The structure of the arithmetical function has vanished.

The objects do not vanish, but rather, there is a transformation in their structural relations. Now P_1 and **S** are related in a particular way to the zero-point in $\odot^\circ$, which is the central (0) object in the structure above. The objects P_1 and **S** are each still a distinct object from (0), but now the distinction no longer exists as a distinction in arithmetical structure. The escape to infinity of n has resulted in the distinction in arithmetical structure transforming into a distinction only in structure space.

Because the arithmetical interval is now zero, the objects P_1 and **S** are now no longer distinct number objects, and are not distinct from the zero-point considered as a number-object (0), but they *are* distinct as mathematical "points". These points, however, are not distinct points in a number-space, specifically, in $\mathbb{C}$, but rather, are distinct points in *structure space*.

In this, is the "secret" of the real continuum $\mathbb{R}$ and the complex continuum $\mathbb{C}$, which each manage to work as a continuum and yet always have distinct and unique points. This working lays in the relationship between fractal infinity and the objects to which the fractal infinity applies, which are the "points", and the number-objects.

The fractal infinity applies to $\mathbb{R}$ and $\mathbb{C}$ through the infinite array in $\odot^{\circ}$, which as we have already seen is also structurally just an effect of $_{\circ}\odot^{\circ}$ to which a zero-point has been applied, and which in turn, itself, is just a mapping possibility arising from $\odot$.

However, the fractal infinity of distinct *points or objects*, is not that same infinity, because we can escape to infinity in an iterative process of "zooming in" on $\mathbb{R}$ or $\mathbb{C}$, either through an arithmetical structure that converges, or through the escaping function $1/(n \to \infty)$, and find that the *arithmetical distinction* between points in $\mathbb{R}$ or $\mathbb{C}$ reduces to zero.

The remaining distinction is in the fractal infinity of structure space $\mathfrak{S}$, which as we have already seen, for mathematical structures is just the infinite instantiations of $\odot$, as distinct objects. This structure *does not reduce* to a zero interval between objects under an arithmetical escape to infinity.

Post naive view of the nature of $\mathbb{R}$, $\mathbb{C}$ and $\mathfrak{S}$.

All this arises because there is a distinction between the structures of mathematical functions, and the nature of structure space, the space of distinct but identical objects. $\mathfrak{S}$ is a conception. But $\mathbb{R}$ and $\mathbb{C}$ *are also* conceptions that happen in our mind, in our own intelligence, the intelligence we are being. If we tacitly envisage this intelligence or mind as an object distinct from the objects $\mathbb{R}$, $\mathbb{C}$, and $\mathfrak{S}$, or *vice versa*, this is just a presumption being made on the basis of our evolutionary experience of the world as a separate and distinct object from our self. However, the actuality is that both our self and the world we experience, as a matter of modern neuroscientific fact, are a construct created through the same principle of the brain.

The most important thing to recognise is the relation between mathematical structures and the mind that conceives them. In particular, that mathematical structures are no more separable from the principle of the brain that arises in nature, than our experience and concept of the world we find ourselves occupying.

Our evolutionary intelligence as it has arisen in nature is in the principle of the brain. Through that, arises our experience of the world, and our conceptions of it. To imagine that these experiences and conceptions are somehow of something separate from, and independent of the principle of the brain, in other words, independent and separate from what modern neuroscience refers to as the "internal model" produced by the principle of the brain, is just naive realism. In order to properly understand nature scientifically, which includes our own brain and intelligence, there is the need to go beyond naive realism both in our outlook and in the interpretation of the mathematics we use in that pursuit.

The Cantor set

There is of course a parallel between the structural function f_M and the process associated with the Cantor set. We remember that in the latter we take the arithmetical interval $0 \rightarrow 1$, and remove the middle third, leaving two intervals $0 \leftrightarrow 1/3$ and $1/3 \leftrightarrow 1$. We then iteratively repeat the process with the remaining intervals, infinitely. The total interval L removed at any iteration number n is well known to be

$$L = \sum_{n=1}^{\infty} \tfrac{1}{2}\left(\tfrac{2}{3}\right)^n$$

which converges to 1. So when $n \rightarrow \infty$ the total interval removed is 1, through a process in which on each iteration only $1/3$ of each remaining interval is ever removed. The remaining, unremoved parts of the intervals, on each iteration, tend to zero as $n \to \infty$. So the situation is that "at infinity" - to use a common term - the entire interval $0 \to 1$ has been removed, and what remains has been reduced to an infinite set of points of zero interval, that are nonetheless separate and distinct, separated by intervals the largest of which is $1/3$.

Simply saying

$$\sum_{n=1}^{\infty} \tfrac{1}{2}\left(\tfrac{2}{3}\right)^n = 1$$

isn't a description of the situation described. The situation or process described is actually a structural function in action.

We can easily specify this process as a structural infinite iteration function $\overset{\infty}{f_D}(0 \leftrightarrow 1)$. Then we are saying that

$$\overset{\infty}{f_D}(0 \leftrightarrow 1) \to (0)$$

Rather than considering the removed intervals as vanishing to nowhere, we can also consider them as being "removed elsewhere", by further specifying that $\overset{\infty}{f_D}$ removes the central thirds on each iteration, and accumulates them in an accumulator A in precisely the same arrangement of relations as they occurred in the original interval $0 \to 1$, so that when $n \to \infty$ $A \equiv (0 \to 1)$.

So over infinite iterations the arithmetical value of A goes from zero to one, and its structure goes from unspecified, to $(0 \to 1)$. The infinite Cantor process - which was already a structural function - together with our additional structural rule, now removes the interval $0 \to 1$ from its former context and puts it in A.

In terms of structures we would say for example

$$\left(0 \xleftrightarrow{\mathbb{R}} 1\right) \xrightarrow{\overset{\infty}{f_D}} (0),$$

$$\left(A \xleftrightarrow{=} 0\right) \xrightarrow{\overset{\infty}{f_D}} \left(A \xleftrightarrow{=} \left(0 \xleftrightarrow{\mathbb{R}} 1\right)\right)$$

We can see here that the infinite Cantor process with our additional rule added to it, does nothing except to change the "location" of the interval $0 \rightarrow 1$. The interval $0 \rightarrow 1$ now in A, is arithmetically one and the same object as the one removed from its former "location". It is composed of the so-called real continuum, as the $0 \rightarrow 1$ interval in $\mathbb{R}$. We merely instantiated the rule and instantiated an object A that we are now saying is the new *location* of that interval. The total interval removed, whose size we worked out from the converging infinite series, is the one that is now in A.

Structure space

What was the interval's original location? Most pure mathematician would not usually regard this as a proper question, because they would normally be interested only in the functional machinery of mathematical structures and how that machinery works, rather than in the question of what these structures *are*, in relation to our own intelligence.

Structure theory acknowledges the location of these structures in our own intelligence, but answers the question in an object-oriented way by saying that the original location was undeclared and unspecified structure space - the space of objects that are distinct, even if they are identical. By instantiating A we instantiated it as a structure space, and through the specification of our structural function we wrote a morphism to turn A into the interval $0 \rightarrow 1$ in $\mathbb{R}$. There was nothing very "structure-like" about this object A to begin with, however. By definition, A is not the same structure space that this interval $0 \rightarrow 1$ was in, to begin with.

The part of the structural function $\overset{\infty}{f}_D$ involving A that we specified, is actually just the inverse of the structural function that the Cantor process itself specifies. But the Cantor process is usually discussed without necessarily recognising it as, or calling it a structural function. Instead of just "removing" thirds from a line or space $0 \rightarrow 1$, $\overset{\infty}{f}_D$ adds them and constructs them together in precisely the opposite way to the Cantor process of removal, to the structural space A.

In order to do so, $\overset{\infty}{f}_D$ begins by first *creating* (in A) the third $1/3 \rightarrow 2/3$. In doing so it has added a zero-point to the structural object A, and specified three intervals in A, and in doing that, it has instantiated $\mathbb{R}$ in A. A was just an instance of $\mathfrak{S}$ that we specified, distinct from the instance of $\mathfrak{S}$ that the interval $0 \to 1$ occupies in the Cantor process. *Any object that we declare, in any circumstance, before further specification, is an instance of $\mathfrak{S}$.*

When we further specify a declared object, then we are introducing new objects, and relating them in $\mathfrak{S}$. Always, the structures we consider are structures in $\mathfrak{S}$ that ultimately have been created *from* $\mathfrak{S}$.

The behaviour of the relation between infinity and numbers is not one that obeys intuition based on tacitly regarding infinity itself as a quantity or quasi-quantity, in some number-space whose origin we haven't even considered. Rather, it derives from the relation between finite *structures* and infinite ones, both kinds of which are created from $\mathfrak{S}$ that is intrinsically infinite. So we have that there is a morphism $\mathfrak{S} \to \mathbf{S}$, through which any structure $\mathfrak{S} \to \mathbf{S}$, finite or infinite, has been created.

There are also morphisms $\mathbf{S} \to \mathfrak{S}$ that occur when there is an escape to infinity, as in the functional division by zero, or the infinite iteration, whether arithmetical or structural. Infinity and zero are themselves related as a structure $0 \leftrightarrow \infty$, where the infinity can be any kind of infinity that you like to conceive or model.

As a conception this infinity is an object in thought, and as an object in thought it is part of a dynamic finite structure whose form and workings can only be understood with evidence-based science, through mathematical structures that are functions of number-spaces. This evidence-based science will be the science of the human brain, and the mathematical structures, will be those that appear in the data that the science uses.

The Object of Nature

Infinity is not an amount, or quantity. As an object in mathematics it manifests either as a process based on iteration, or as the escape of an object that was previously a number-object, from identification as a number.

The *native* object that escapes or doesn't escape, is not the number-object, or the algebraic object used to represent the number-object, but an object we have been representing as a structure in $\mathfrak{S}$. The creation of number-objects requires first the inception of the zero-point in $\odot$, which instantiates $\mathfrak{S}$, where $\mathfrak{S}$ itself arises *as* the possibility of mappings in $\odot$, as $_{\circ}\odot^{\circ}$ and $\odot^{\circ}$, and then number-relations.

So the creation of a *structure* requires the instantiation of a zero-point in $\odot$, whereupon numbers and relations between them can arise. This is just a pure mathematical description of the creation of a *structure*. It implies "nothing" as the starting point for "something", where mathematically, "nothing" is the zero number, but structurally, the zero number begins its life, so to speak, as a "zero-point" devoid of any "number" inference, arising on a bounded infinity (which in itself is not a number, and doesn't consist of Number). The bounded infinity, in itself, as the primary object, is also devoid of any number inference before the imposition of the "zero-point".

So in terms of these structures, in an object-oriented way, "something" can arise out of "nothing", if "nothing" itself is identified as a bounded infinity object, and the arising of "something" is identified as the arising of a "zero-point", from which spontaneously springs, so to speak, structure space.

Imagine, if you will, the object of nature. The object of the whole of nature, the whole of the universe, or multiverse, or whatever you conceive it as. The word *universe* will always do, because its literal meaning is "everything turned into one". But because it is not only stars and galaxies and so forth, that astronomers refer to as the universe, then the word *nature* will do best. So imagine, if you will, the object of nature.

What does it look like? It can't look like anything, because all looking and seeing is part of the sentient intelligence we are being, that has arisen in nature. Nature comes first, and then the sentient intelligence we are being. Not first in time, though, because time itself is something that arises in nature.

So we are going to propose that the object of nature, as an object, is nothing. But we are also going to propose that this nothing instantiates itself infinitely, in infinitely many ways, all of them unique. There is no mathematical equivalent of this, because the mathematical nothing is zero, which isn't truly nothing. It is a number, and a number is a part of the human intelligence we are being. Rather, we are now speaking of nothing.

Let us now propose in our play of conception that this is instantiated infinitely, by virtue of itself, and that this infinity of instantiation is what now constitutes the *something*. From the point of view of the intelligence we are being, the one that has evolved in time in nature, our given brain-produced intelligence can't expect to see what this something *is*. Just as it can't know what the nothing that it arises from, *is*. But there is now an infinity of relations between these instantiations. We could even say this is a structure. And then we can also say there will be infinite substructures.

If in object-oriented conception we happen to conceive in our intelligence, as it stands, the nothing as bounded infinity, then the possibility of all mathematical structures and relations arises, because structure space arises.

We can then begin to conceive, in our naturally created and given intelligence, an object-oriented model of the play of structures and relations. This play arises in a vast emergence, arising in what we as the brain-produced intelligence we are conditionally being, know as time, space, matter, and self, in other words, the intelligence we are conditionally being.

www.ingramcontent.com/pod-product-compliance
Lightning Source LLC
Chambersburg PA
CBHW081544220326
41598CB00036B/6554